11/95

D0436495

ON THE SHOULDERS OF GIANTS

THE HISTORY OF SCIENCE FROM 1946 TO THE 1990s

Ray Spangenburg and Diane K. Moser

Facts On File®

AN INFOBASE HOLDINGS COMPANY

On the cover: Chien-shiung Wu, professor of physics at Columbia University (UPI/Bettman)

The History of Science from 1945 to the 1990s

Copyright © 1994 by Ray Spangenburg and Diane K. Moser

All rights reserved. No part of this book may be reproduced or utilized in any form or by any means, electronic or mechanical, including photocopying, recording, or by any information storage or retrieval systems, without permission in writing from the publisher. For information contact:

Facts On File, Inc.
460 Park Avenue South
New York NY 10016

Spangenburg, Ray, 1939–
    The history of science from 1946 to the 1990s / Ray Spangenburg
and Diane K. Moser.
        p. cm — (On the shoulders of giants)
    Includes bibliographical references and index.
    ISBN 0–8160–2743–9
    1. Science—History—Juvenile literature. 2. Life sciences—
History—Juvenile literature. [1. Science—History.] I. Moser,
Diane K., 1944–   . II. Title. III. Series: Spangenburg, Ray,
1939–   On the shoulders of giants.
Q126.4.S63 1994
5091 .045—dc20   93–46058

Facts On File books are available at special discounts when purchased in bulk quantities for businesses, associations, institutions or sales promotions. Please call our Special Sales Department in New York at 212/683–2244 or 800/322–8755.

Text design by Ron Monteleone/Layout by Robert Yaffe
Cover design by Semadar Megged
Printed in the United States of America

MP FOF 10 9 8 7 6 5 4 3 2 1

This book is printed on acid-free paper.

509
SPA
1993
v.5

———————

To a certain long-haired mutt,
for having put up with many months
of long, boring dawn-to-midnight days
with not nearly as many games of
chase-the-stolen-washcloth

or

tug-a-toy
as he would have liked

C O N T E N T S

ACKNOWLEDGMENTS

Books are a team effort, and this one is no exception. So many people have been generous with their time, talents and expertise in helping us with this book, both as we wrote it and in the past when it was only a dream—more than we can possibly name. We'd like to thank everyone for their help and encouragement, and a special thank you to: Gregg Proctor and the rest of the staff at the branches of the Sacramento Public Library for their tireless help in locating research materials. Shawn Carlson, of Lawrence Berkeley Laboratories; David Deamer of UC Davis; Richard Ingraham, formerly of California Sate University at San Jose; as well as Beth Etgen, educational director at the Sacramento Science Center, and her staff for kindly reading parts of the manuscript and making many helpful suggestions. Andrew Fraknoi of the Astronomical Society of the Pacific, for his help with research contacts and illustrations. Also for their help with illustrations: Doug Egan of the Emilio Segrè Visual Archives, Pat Orr at the California Institute of Technology and Yae Shinomiya of Stanford University. Thanks above all to Facts On File's editorial team, especially Nicole Bowen for her intelligent, energetic and up-beat management of the project; Janet McDonald for her eagle-eyed passion for detail; Christa Sidman, for helping us stay on top of production details; and James Warren for his vision and encouragement in getting it started. And to many others, including Jeanne Sheldon-Parsons, Laurie Wise, Chris McKay of NASA Ames, Robert Sheaffer and Bob Steiner, for many long conversations about science, its history and its purpose.

P R O L O G U E

THE SCIENTIFIC METHOD

. . . our eyes once opened, . . . we can never go back to the old outlook. . . . But in each revolution of scientific thought new words are set to the old music, and that which has gone before is not destroyed but refocused.
—A. S. Eddington

What is science? How is it different from other ways of thinking? And what are scientists like? How do they think and what do they mean when they talk about "doing science?"

Science isn't just test tubes or strange apparatus. And it's not just frog dissections or names of plant species. Science is a way of thinking, a vital, ever-growing way of looking at the world. It is a way of discovering how the

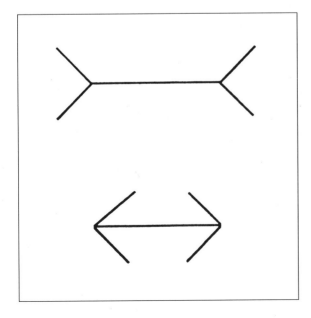

Looks can be deceiving: These two lines are the same length.

world works—a very particular way that uses a set of rules devised by scientists to help them also discover their own mistakes.

Everyone knows how easy it is to make a mistake about the things you see or hear or perceive in any way. If you don't believe it, look at the two horizontal lines on the previous page. One looks like a two-way arrow; the other has the arrow heads inverted. Which one do you think is longer (not including the "arrow heads")? Now measure them both. Right: They are exactly the same length. Because it's so easy to go wrong in making observations and drawing conclusions, people developed a system, a "scientific method," for asking "How can I be sure?" If you actually took the time to measure the two lines in our example, instead of just taking our word that both lines are the same length, then you were thinking like a scientist. You were testing your own observation. You were testing the information that we had given you that both lines "are exactly the same length." And, you were employing one of the strongest tools of science to do your test: You were quantifying, or measuring, the lines.

Some 2,400 years ago, the Greek philosopher Aristotle told the world that when two objects of different weights were dropped from a height the heaviest would hit the ground first. It was a common-sense argument. After all, anyone who wanted to try a test could make an "observation" and see that if you dropped a leaf and a stone together the stone would land first. Try it yourself with a sheet of notebook paper and a paperweight in your living room. Not many Greeks tried such a test, though. Why bother when the answer was already known? And, being philosophers who believed in the power of the human mind to simply "reason" such things out without having to resort to "tests," they considered such an activity to be intellectually and socially unacceptable.

Centuries later, Galileo Galilei, a brilliant Italian who liked to figure things out for himself, did run some tests. Galileo, like today's scientists, wasn't content merely to observe the objects falling. Using two balls of different weights, a time- keeping device, and an inclined plane, or ramp, he allowed the balls to roll down the ramp and carefully *measured* the time it took. And, he did this not once, but many times, inclining planes at many different angles. His results, which still offend the common sense of many people today, demonstrated that, if you discount air resistance, all objects would hit the ground at the same time. In a perfect vacuum (which couldn't be created in Galileo's time), all objects released at the same time from the same height would fall at the same rate! You can run a rough test of this yourself (although it is by no means a really accurate experiment), by crumpling the notebook paper into a ball and then dropping it at the same time as the paperweight.

Galileo's experiments (which he carefully recorded step by step) and his conclusions based on these experiments demonstrate another important attribute of science. Anyone who wanted to could duplicate the experiments and either verify his results or, by looking for flaws or errors in the experiments, prove him partially or wholly incorrect. No one ever proved Galileo wrong. And years later when it was possible to create a vacuum (even though his experiments had been accurate enough to win everybody over long before that), his conclusions passed the test.

Galileo had not only shown that Aristotle had been wrong. He demonstrated how, by observation, experiment and quantification, Aristotle, if he had so wished, might have proven himself wrong—and thus changed his own opinion! Above all else the scientific way of thinking is a way to keep yourself from fooling yourself—or, from letting nature (or others) fool you.

Of course science is more than observation, experimentation and presentation of results. No one today can read a newspaper or a magazine without becoming quickly aware of the fact that science is always bubbling with "theories." "ASTRONOMER AT X OBSERVATORY HAS FOUND STARTLING NEW EVIDENCE THAT THROWS INTO QUESTION EINSTEIN'S THEORY OF RELATIVITY," says a magazine. "SCHOOL SYSTEM IN THE STATE OF Y CONDEMNS BOOKS THAT UNQUESTIONINGLY ACCEPT DARWIN'S THEORY OF EVOLUTION," proclaims a newspaper. "BIZARRE NEW RESULTS IN QUANTUM THEORY SAY THAT YOU MAY NOT EXIST!" shouts another paper. What is this thing called "theory?"

Few scientists pretend anymore that they make use of the completely "detached" and objective "scientific method" proposed by the philosopher Francis Bacon and others at the dawn of the scientific revolution in the 17th century. This "method," in its simplest form, proposed that in attempting to answer the questions put forward by nature, the investigator into nature's secrets must objectively and without preformed opinions observe, experiment and gather data about the phenomena. After Isaac Newton demonstrated the universal law of gravity, some curious thinkers suggested that he might have an idea *what gravity was.* But he did not see such speculation as part of his role as a scientist. "I make no hypotheses," he asserted firmly. Historians have noted that Newton apparently did have a couple of ideas, or "hypotheses," as to the possible nature of gravity, but for the most part he kept these private. As far as Newton was concerned there had already been enough hypothesizing and too little attention paid to the careful gathering of testable facts and figures.

Today, though, we know that scientists don't always follow along the simple and neat pathways laid out by the trail guide called the "scientific method." Sometimes, either before or after experiments, a scientist will get an idea or a hunch (that is, a somewhat less than well thought out hypothesis) that suggests a new approach or a different way of looking at a problem. Then he or she will run experiments and gather data to attempt to prove or disprove this hypothesis. Sometimes the word hypothesis is used more loosely in everyday conversation, but for a hypothesis to be valid scientifically it must have built within it some way that it can be proven wrong, if, in fact, it is wrong. That is, it must be falsifiable.

Not all scientists actually run experiments themselves. Most theoreticians, for instance, map out their arguments mathematically. But hypotheses, to be taken seriously by the scientific community, must always carry with them the seeds of falsifiability by experiment and observation.

To become a theory, a hypothesis has to pass several tests. It has to hold up under experiments, not only to the scientist conducting the experiments or making the observations, but to others performing other experiments and observations as well. Then when thoroughly reinforced by continual testing and appraising, the hypothesis may become known to the scientific or popular world as a "theory."

It's important to remember that even a theory is also subject to falsification or correction. A good theory, for instance, will make "predictions"—events that its testers can look for as a further test of its validity. By the time most well-known theories such as Einstein's theory of relativity or Darwin's theory of evolution reach the textbook stage, they have survived the gamut of verification to the extent that they have become productive working tools for other scientists. But in science, no theory can be accepted as completely "proven"; it must remain always open to further tests and scrutiny as new facts or observations emerge. It is this insistently self-correcting nature of science that makes it both the most demanding and the most productive of humankind's attempts to understand the workings of nature. This kind of critical thinking is the key element of doing science.

The cartoon-version scientist portrayed as a bespectacled, rigid man in a white coat, certain of his own infallibility, couldn't be farther from reality. Scientists, both men and women, are as human as the rest of us—and they come in all races, sizes and appearances, with and without eyeglasses. As a group, because their methodology focuses so specifically on fallibility and critical thinking, they are probably even more aware than the rest of us of how easy it is to be wrong. But they like being right whenever possible and they like working toward finding the right answers to questions. That's usually why they became scientists.

INTRODUCTION

FROM 1946 TO THE 1990s:
THE CONTINUING QUEST

The year 1945 was both a bitter and jubilant time. A terrible war had finally ground to an end—a world war that had annihilated millions of Jews, destroyed much of Europe, unhinged large parts of Asia and the Pacific, decimated two cities of Japan and killed millions of other soldiers and civilians. At last, with great relief, the world could get back to the business of living. And scientists could get back to doing science.

But the political ambience in which they worked was not peaceful, and the world psyche after World War II was deeply shaken. The atomic bomb, with its enormous destructive capacity and its deadly radiation aftermath, had been unleashed, and the world could not escape the new anxieties created by the bomb's existence. By 1949, the Soviet Union, with its declared aggressive policies, had tested an atomic bomb, and a race began between the United States and the Soviet Union to see who could stockpile the most arms. In the United States, some people began building bomb shelters in their backyards, and air raid drills remained part of every public school routine throughout the 1950s. The Soviet Union established what became called an Iron Curtain in eastern Europe, a closed-door policy of noncooperation, whereby freedom of travel and economic exchange were prohibited. So, while World War II had finally ended, a new kind of conflict, known as the Cold War, began almost immediately.

In 1950 war broke out when North Korean troops, bolstered by China, invaded South Korea, which was defended by United Nations troops. The 1950s and 1960s were a time of many civil wars and coups in Latin America and Africa, in many cases encouraged either by the Soviet Union or the United States. Throughout the second half-century, independence was won, for the most part bloodlessly, by former colonies in Africa, Asia and Latin America. But many—such as Cuba, Tibet, North Korea, and several Latin American countries—replaced colonial rule with commitments to the Soviet Union or Communist China. And in 1962, the whole world caught its breath as the United States confronted the Soviet Union with evidence of Soviet missile bases in Cuba, only about 90 miles from the coast of

Florida—and breathed a sigh of relief when the Soviet Union backed down. Just a few months later, the two governments installed a communications "hot line" between Moscow and Washington, D.C. to reduce the risk of accidental war between the two nations.

For science, this pervasive atmosphere of unease had both positive and negative effects. The free exchange of ideas, theories and results among scientists—so hard fought for over the centuries—now fell by the wayside as communication fell silent between Moscow, its satellites and the West. Soviet newspapers, journals and books were not available in the West, and vice versa. Travel between the two regions was severely limited. But, on the positive side, the United States, Europe and England early recognized the need to keep apace with Soviet advances. And in the United States especially, the tradition begun by the scientists who built the atomic bomb during the Manhattan Project was continued after the war by scientists who coupled their search for scientific knowledge with the need to build advanced military aircraft and weapons for their governments.

Then science in the West received a huge jolt. On October 4, 1957 the Russians launched *Sputnik I,* the first artificial satellite, into orbit around the Earth. No other countries were close to launching anything into orbit, except the United States, where rocket scientists imported from Germany after World War II had been working steadily both on military missiles and on the raw beginnings of a civilian space program. Suddenly math and science moved center stage. Spurred by competition with the Russians, Western educators put new emphasis on grooming young scientists, while in the United States, the precursor of the National Aeronautics and Space Administration (NASA) scrambled to catch up. By 1961, both the Soviet Union and the United States had sent more than one human into space (with the Russians again first), and by 1961, U.S. president John F. Kennedy had announced plans to send Americans to land on the Moon. To the arms race the two nations now added a race to the Moon, and the Space Age had begun in earnest.

For planetary scientists and astronomers, the boon was enormous. Between 1958 and 1976, scientists sent 80 missions to the Moon, 8 of them manned. From these missions, scientists obtained vast amounts of new information, including the first photographs of the Moon's far side (which always turns away from the Earth). Probes to Venus, Mars, Mercury and the far-flung outer planets followed, both from Cape Canaveral (later briefly renamed Cape Kennedy) and Russia's "Star City." These roaming robots sent radio signals earthward, returning photographs and data that revolutionized human understanding of the Solar System and the universe beyond.

By 1971, the Soviet Union had begun a marathon of experiments in living in space, in a series of space stations culminating in the big Mir space station, launched in 1986. The United States, meanwhile, sent three teams of astronauts in 1973–74 to a space station called Skylab to study the Sun. And in 1982, the United States launched the first of a fleet of Earth-orbiting vehicles called the Space Shuttles. In addition to deployment of countless satellites having a variety of purposes ranging from scientific to military to business, the space programs of these two countries have collected extensive data about the effects of weightlessness on living organisms (including humans), crystal formation, and countless other areas of inquiry. By the 1980s, many other countries, including China, Japan and India, had developed the capability to launch satellites and had developed space programs of their own.

All this was, in a way, positive fallout from the Cold War. As the Cold War drew toward an end in 1989 and with the breakup of the Soviet Union in 1991, some of the spur for this surge of interest in space exploration and science dissipated. The Mir space station has been limping along in the early 1990s, and the U.S. Congress reduced funds for an international space station called Freedom, not yet built. As Russia and its neighbors struggled to build new economies and new governments and to settle age-old differences, the pursuit of science began to take a back seat for a while.

At the same time, scientists throughout the world have become excited about their ability to communicate and collaborate freely with scientists from Russia and other former Iron Curtain countries. Ironically, the new flow of information has come at a time when funds are dwindling.

And if any single fact of life has haunted science in the last half of the 20th century it is the greatly increased cost of doing science. Gone are the days when Galileo, in the 17th century, could hear about a new optical instrument, gather together a few materials, build his own telescope and shortly be making astronomical observations that no one had ever made before. With 1945 dawned the age of Big Science. Particle physics, the study of subatomic particles, could only be conducted with the aid of giant machines, called particle accelerators, that began to be built, one by one, in Stanford and Berkeley, California; in Batavia, Illinois; in Geneva, Switzerland; and elsewhere. As Europe struggled to recover from the devastation of the war, most of the early work was done in the United States, and physicists flocked there from Japan, China and Europe.

Only well-endowed universities and institutions, with the help of government funding, could afford to build the big machines. The giant computer and semiconductor industry, as well, spun off from advances made by physicists and engineers—discoveries about the behavior of electrons and quantum mechanics; the discovery of the transistor by a team of scientists at Bell Laboratories (again, through research funded by a large corporation, not the individual enterprise of a single curious mind). And the first commercial computer to go into service was the UNIVAC I, a giant machine built in 1951 and purchased by the U.S. Bureau of the Census. Again, only the enormous needs and financial capabilities of government could justify and finance the start-up. (Today, for under $600, any individual can buy a personal computer the size of a briefcase with greater capacity than the UNIVAC I, which filled an entire room the size of a gym.)

The period from 1945 to the present has been a time of turbulence and change, a time both of violence and social progress.

Civil rights and liberties have become a matter of worldwide concern, and some progress was made in the United States under the leadership of Martin Luther King, Jr. and others in the 1960s—but not without high costs. Martin Luther King, Jr. was assassinated in 1968. The U.S. Supreme Court ordered integration of schools in 1954 and federal troops enforced it in Little Rock, Arkansas in 1957. In the 1960s civil rights legislation in the United States established the principle of equal opportunity for all in jobs and housing. And by 1991, even South Africa had abandoned apartheid laws and had begun to integrate schools.

Political assassinations also haunted the times. President John F. Kennedy and his brother, Robert F. Kennedy, were assassinated—John in Dallas, Texas in 1963 and his brother on the night he won the California presidential primary four and a half years later, in 1968. In India, pacifist leader Mahatma Gandhi, who had led India successfully in a fight for independence from Britain, was assassinated by a fanatic in 1948. Extremists struck again, nearly 40 years later, in 1984, assassinating Prime Minister Indira Gandhi (unrelated to Mahatma Gandhi), and seven years later, her son, Rajiv Gandhi, who had succeeded her, was also killed. Other assassinations, terrorist bombings and hijackings worldwide reflected an atmosphere of turbulence. But at the same time, forces for peace, national independence and self-rule often triumphed as well.

The fight for democratic process and self-rule had its ups and downs, with several significant changes in addition to the breakup of the Soviet Union. In 1986, Corazón Aquino was elected president of the Philippines, putting an end to 15 years of martial law and 20 years of rule by the graft-ridden government of Ferdinand Marcos. East Germany and West Germany were reunited in 1990 and held the first democratic elections in a unified Germany since 1932.

Science, meanwhile, could not stand apart from these political, social and moral issues. Many scientists took a stand after World War II against the continued development of weapons, among them Albert Einstein, whose sphere of influence is legendary, and Danish physicist Niels Bohr. Andrei Sakharov, in Russia, who helped his nation develop a hydrogen bomb, spoke out in 1967–68 against testing of nuclear weapons in the Soviet Union and for worldwide disarmament—an act for which he was unjustly discredited, persecuted and threatened. Finally, in 1980 when he criticized the Soviet Union for its invasion of Afghanistan, he and his wife, Elena Bonner, were placed under house arrest. There he remained imprisoned until 1986, when the growing spirit of *glasnost,* or openness, led to his release.

As expanding knowledge creates new areas of concern for ethics and new decisions to be made, science walks close to the heart and soul of society on many issues. Is a cloned tomato better than a natural tomato, or should it be suspect in some way? When donor organs can save lives, when and under what circumstances can they be taken from the donor? Given the nuclear accidents at Three-Mile Island (in Pennsylvania) in 1979 and at Chernobyl (near Kiev, in the Soviet Union) in 1986, can nuclear power plants be considered safe?

Science is often viewed as both hero and villain in the second half of the 20th century. While science has enabled enormous technological advances—from electricity to compact discs, from automobiles to airplanes to space exploration, from satellite communication to fax machines—science also sometimes gets blamed for the loss of the "simple, natural life." But the cycle continues: discovery spawning new technology, which in turn makes new discoveries possible, in a series of leap-frog jumps into the future. The process requires new understanding of the effects of what we do, new attitudes of stewardship toward our planet and a new sense of responsibility for the effects of our actions upon our neighbors—responsibilities that our forebears, with their "simpler life" of timber harvesting and poor crop planning, often failed.

The second half of the 20th century has been a time for exploring the most fundamental components of the universe, the essence of which things are made. Leucippus and Democritus, among the Greeks, believed that all matter was composed of atoms, which they imagined to be tiny, hard, indivisible particles. In the 19th century, John Dalton believed he knew what an atom was: the smallest unit of a chemical element. But in the last years of the 1800s, chemists and physicists such as Marie and Pierre Curie and Henri Becquerel noticed that the atoms of certain elements seemed to give off a part of themselves in a process that we now call radioactive decay. They asked themselves, if an atom was indivisible, how could it emit part of itself? Here was a clear contradiction, and in the first years of the 20th century the stage was set for revolutionary changes in the way scientists understood the atom. Electrons were discovered, followed by a nucleus composed of protons and neutrons.

But by 1945, exploration of the new world within the atom had only just begun. Today some 200 subatomic particles are known, and more are believed to exist. The story of their discovery has been an intricate and intriguing whodunit that has absorbed some of the best minds of the century.

Meanwhile, enormous technological breakthroughs in rocket technology and space science have enabled astronomers, cosmologists and planetary scientists to explore the universe more closely than Galileo, Kepler or any of the great observers of the past could ever have dreamed possible.

At the same time, in the life sciences, researchers were eagerly racing to discover the structure of the fundamental constituents that control the shapes and forms of life. By 1946, it had been clear for two years that the ultimate architect of living things was a molecule known as deoxyribonucleic acid, or DNA—a substance that had been completely unknown before the first half of the 20th century. And its structure still remained a mystery as the second half of the century began. But it would soon be unraveled by James Watson and Francis Crick, and the new field of molecular biology would become the cutting edge of the life sciences for the rest of the century.

Today, there are as many scientists currently at work in the world as have existed in the entire history of science. In 1987, some 264,900 students received doctorates in science or engineering from U.S. universities. And more than $103.9 billion were spent on research and development in the United States in 1988, as compared, in constant 1982 dollars, with only $62.4 billion in 1970. Hundreds of specialties from computer science to microbiology and from astrophysics to particle physics attract young people who have a thirst for knowledge.

In this short book, it is possible only to explore *some* of the exciting discoveries made in the second half of the 20th century—in light of the explosion of new finds and intriguing questions raised every day. At the end of this book, we've suggested several books for further exploration, as well as other resources, such as magazines and organizations. We hope you'll plunge into these, too, and discover for yourself some of the fascinating complexities and delightful questions our universe poses. For science provides us all, whether we are professional scientists or not, with a special window on the world that enables us to see in ways we might not otherwise see and understand in ways we would not otherwise understand. It is a special, and uniquely human, way of thinking.

This book, like the four others in the series called *On the Shoulders of Giants*, looks at how people have developed a system for finding out how the world works, making use of both success and failure. In it, we will look at the theories they put forth, sometimes right and sometimes wrong. And we'll look at how we have learned to test, accept and build upon those theories—or to correct, expand or simplify them.

We'll also see how scientists have learned from others' mistakes, sometimes having to discard theories that once seemed logical but later proved to be incorrect, misleading or unfruitful. In all these ways they have built upon the shoulders of other men and women of science—the giants who went before them.

THE PHYSICAL SCIENCES FROM 1946 TO THE 1990s

THE SUBATOMIC WORLD:
A VAST SWARM OF PARTICLES

*B*eneath a long, curving overpass of Interstate 280 on the San Francisco peninsula, a four-mile-long building cuts through the gently
rolling Los Gatos Hills. Most people don't even see the unimportant-
looking brown concrete slab, or if they do, they have no sense, as they
whiz overhead, of its extraordinary length or the millions of high-speed
electrons that speed through its tunnels at a pace that makes the vehicles
on the freeway look like they're traveling in mega-slow-motion. This
great gash of brown concrete across the manzanita brush and grassy
slopes doesn't really look remarkable, even to those who do notice it,
except that it's a great deal longer than normal buildings and strangely
lacking in corners or bends.

But it is remarkable. This building, known casually to its friends as
"SLAC," houses one of modern physicists' answers to an age-old question:
How do you see the smallest conceivable constituents of matter—the tiny
building blocks of which everything is made? SLAC (the Stanford Linear
Accelerator Center), huge as it is, provides a window for seeing the infini-
tesimally tiny particles—known as elementary particles—that make up the
atom.

BEGINNING THE SEARCH

But let's back up a little. At the outset (or close to it, as far as the history of
science is concerned), back in the time of the ancient Greeks, a man named
Leucippus [lyoo SIP us] and his disciple, Democritus [de MOK rih tus] came
up with the idea that everything must be made of some very small universal
building block, and they called this tiny, hard, indivisible particle an *atom*

*Stanford Linear
Accelerator Center*
(Stanford University
News Service)

(from the Greek word *atomos*, meaning "indivisible," or, put another way, "unsmashable"). Too small to see, these atoms were what you would get if you kept splitting any material into its component parts until it was impossible to split it any further.

That was about 2,400 years ago, and the idea caught on slowly. Most other thinkers of their time and for many centuries afterward didn't think much of the idea, but by the late 1600s, it began to attract more interest. The British chemist Robert Boyle was an atomist, and Isaac Newton also liked the idea. In *Opticks*, published in 1704, Newton wrote that he believed all matter was composed of "solid, massy, hard, impenetrable moveable Particles" that he thought must be solids "incomparably harder than any porous Bodies compounded of them." But even Newton had no idea how to look at these particles he thought must exist. So physicists went on studying the relationships of energy and matter, cause and effect. Chemists, meanwhile, continued searching for what came to be called "elements."

A little less than a century later, an eccentric, headstrong chemist named John Dalton proposed, for the first time, an atomic theory that could be tested quantitatively. Dalton defined atoms as the smallest unit of an element, and he published the first list of atomic weights for the known elements of his time. His work provided the keystone for several dozen discoveries of new elements in the century that followed. But what Dalton and his contemporaries didn't yet realize was that what Dalton called an "atom" was *not* Leucippus and Democritus's indivisible "atom," the smallest fundamental particle to be found in nature.

The first clues that this was not so began coming in at the end of the 19th century with the discovery of X rays and other forms of radiation. This radiation, it turned out, was composed of particles given off by atoms. If atoms could give off particles, then they were not indivisible: There was *something* smaller. In 1897, the English physicist J. J. Thomson demonstrated the existence of the electron, a small, light, negatively charged particle that had a mass equal to only a small fraction of the mass of hydrogen atoms. This didn't explain where radioactive particles came from, but it was a start. And so began the field of subatomic particle physics.

In 1911, New Zealander Ernest Rutherford concluded, from experiments he and his research teams did at McGill University in Canada and at Manchester University in England, that the atom was mostly empty space, with Thomson's negatively charged electrons in orbit around the perimeter, like tiny planets in a miniature solar system, and a nucleus at its center, composed of positive particles (soon named *protons*).

Danish physicist Niels Bohr, who arrived in England to study in 1912, was one of the few physicists who liked this idea of an atom composed mostly of empty space. By 1913, he proposed an updated version of Rutherford's model of the atom that assumed a central positive nucleus surrounded by electrons orbiting at various specific energy levels. Bohr's model combined Thomson's electrons and Rutherford's experimental evidence for the existence of a positive nucleus with the theory of the quantum first proposed by Max Planck in 1900. The basic idea behind Planck's theory, as it began to evolve, is that you can explain how atoms behave—and how subatomic particles interact—based on seeing the photon (the tiny packet of energy that light and all electromagnetic energy travel in) as both a wave and a particle, not one or the other. While this idea seems strange, the theory of the quantum explains a great many phenomena that otherwise had no explanation, and in the end revolutionized physics.

By the third decade of the 20th century rapid-fire changes began to occur in physics; one after another new particles were discovered, and with each new discovery another revision had to be made of ideas about what an atom was like. The new view of Dalton's atom soon bore no resemblance at all to Leucippus and Democritus's unsplittable, billiard-ball-shaped fundamental particle.

In 1930, based on his study of experimental data, Wolfgang Pauli came up with the idea that a strange unknown particle must be given off in radioactive beta emissions, a particle having no mass (or nearly so), no charge and practically no interaction with anything else. He thought such a particle must exist in order to explain the loss of energy in the reaction— otherwise the law of conservation of energy would have to be abandoned, a move he didn't want to suggest. Four years later, Enrico Fermi gave his stamp of approval by naming the tiny particle *neutrino*, which means "little neutral one" in Italian, Fermi's native language.

Almost impossible to detect, the neutrino eluded detection for years, and no one could prove that it existed. There were those, at first, who suspected that Pauli had just pulled a sort of bookkeeping trick—made something up to make the energy books look balanced. But in 1956, through the use of a nuclear power station, an elaborate experiment was completed that proved the existence of the ghostly neutrino, and Pauli was vindicated.

Also in 1930, from a theory proposed by Paul Dirac, a 28-year-old British physicist, came the idea of another hypothetical particle, similar to an electron, but having a positive charge. In fact, based on Dirac's efforts to get quantum and relativity to work together, physicists came to the astonishing conclusion that wherever matter existed, its mirror must also exist, which became known as antimatter. This idea of antimatter, according to fellow Nobel-winning physicist Werner Heisenberg, was "perhaps the biggest of all the big jumps in physics in our century." The idea met with some opposition, despite Dirac's mathematical prowess. Then, in 1932, a young physicist named Carl Anderson was working with a powerful magnet and a cloud chamber at the California Institute of Technology, when he saw it—at least he saw the tracks of a subatomic particle that seemed to be an electron, but was pulled the opposite way by the magnet. He called his new particle a *positron*.

Also in 1932, meanwhile, James Chadwick at Cambridge University found undeniable evidence of the existence of yet another uncharged particle, this one located in the nucleus of most atoms. He called it the *neutron*. This particle was easily detectable and it explained a lot, including the discrepancy that no one had ever understood between atomic number and atomic weight. The number of negatively charged electrons balanced the number of positively charged protons in the nucleus, but all atoms except hydrogen had at least twice as much mass than these particles could account for. So where did all the mass come from? Now the answer seemed clear: neutrally charged particles in the nucleus.

Hideki Yukawa (1907–81) set physicists on a long and fruitful search for fundamental particles (AIP Emilio Segrè Visual Archives, W. F. Meggers Collection)

The existence of neutrons seemed readily substantiated, and it was hard to argue with Carl Anderson's evidence for the positron. A few scientists at this stage were still skeptical about Pauli's neutrino. But, in any case, even with these additions, the number of fundamental particles within the atom still seemed manageable.

But in the coming years all that was about to change. In 1935 Hideki Yukawa (1907–81), a young Japanese physicist at the University of Kyoto, proposed a solution to a major problem: What held these neutrons and protons so closely together in the nucleus? If the nucleus contained only positive protons and Chadwick's chargeless neutrons, Heisenberg had suggested, then the only electrical charge in the nucleus was positive. And since like-charged particles repel each other, why didn't the tiny particles fly apart in opposite directions? Maybe, suggested Yukawa, some sort of "exchange forces" were at work in the nucleus—but what they were and how they worked he never said.

Yukawa figured that, since ordinary electromagnetic forces involved the transfer of photons, then some kind of "nuclear force" must operate within the nucleus, involving the transfer of some other entity. This nuclear force must operate only over very short ranges—ranges, that is, the size of the diameter of a nucleus (measuring about a ten-trillionth of a centimeter). It must also be remarkably strong—strong enough to hold together protons whose positive charges would ordinarily force them apart. And, based on the results of experiments, this force must drop off very quickly as distance increases, so that beyond the perimeter of the nucleus, it would not be felt at all.

Yukawa worked out a theory that evidence of this force appeared when particles were transferred back and forth among the neutrons and protons. The mass of these particles, he said, would depend on the range over which the force operated. The shorter the range, the greater the mass would have to be. To function across the width of a nucleus, transported particles would have to be about 200 times the mass of an electron and about one-ninth the mass of a neutron or proton.

The short-range force that carried these particles came to be called, logically enough, the *strong* force. As for Yukawa's particle, for several years it was known as the *Yukon* in his honor, but the name *meson* is the one that stuck, referring to the fact that mesons, as understood at the time, had a mass intermediate between the proton's and the electron's. (More recently, it has become known as the "Yukawa Particle.") Lo and behold, the following year, using the very same equipment he had used to track the telltale path of the positron, Carl Anderson found evidence of what he thought was the meson. As it later turned out, Anderson's new particle was not a meson but another particle, called the muon, and Yukawa's was not confirmed until 1947.

By 1947 the list of ultimate building blocks of matter and radiation had expanded to include the electron and its twin, the positron, the proton, the neutron, the muon, the pion, the neutrino and the photon. As it turned out, these were not all as elementary as physicists thought at the time; before long, they would discover that protons, neutrons and mesons all broke down into even smaller constituents. Yukawa had set physicists on a search for the small and fundamental that would soon expand the number of subatomic particles to the hundreds. Dalton would have been amazed to see his ultimate fundamental particle now.

And so began our entrance into the subatomic world—bizarre, absurd and astounding. Before long, physicists had a new inventory of extraordinary names to describe the minuscule world of subatomic particles and the forces at work there. They began to talk of tiny particles with whimsical-sounding names like *lepton, muon, pion, gluon,* and, most fantastic of all, *quark*—which they discussed in terms of *strangeness, charm, colors* and *flavors*.

TRIUMPH OF THE CHIP
AND THE DIGITAL COMPUTER

In 1948, three researchers at Bell Laboratories in New Jersey made an amazing discovery. William Bradford Shockley, John Bardeen and Walter Brattain found that they could use certain impure crystals to work the same way Thomas Alva Edison had used vacuum tubes: as electronic devices, or transistors, that control the flow of electrons. And so solid-state semi-conductor electronics was born. The result has affected nearly every aspect of our lives, from the kitchen to automobiles to business, communication and space. Immediately, radios were transformed from large pieces of furniture that people gathered around in their living rooms to little boxes that could be carried everywhere they went. Soon television sets got small enough to slip into a shirt pocket. But one of the biggest influences was on the giant digital computers that had begun to crunch numbers for institutions such as the U.S. Bureau of the Census, banks and big corporations. Digital computing had found a place for itself by the 1940s in the handling of records wherein millions of transactions had to be tracked and accessed. But vacuum-tube computers used vast amounts of energy, required huge, air-conditioned rooms and were notoriously unreliable—going "down," it seemed, more often than they were operational.

The reliable descendant of the transistor—the microchip—soon made computers both dramatically cheaper and incredibly smaller. Today, for under a thousand dollars, people buy computers to set on a desk at home

QUANTUM JOINED TO MAXWELL'S THEORY

During the war, between 1941 and 1945, the United States had gathered together at Los Alamos in north central New Mexico the most prodigious group of physicists ever gathered for a project. Their objective: to build an atomic—or, more accurately, a nuclear fission—bomb before physicists in Nazi Germany could do the same. The Manhattan Project, as it was called, succeeded, making its first test in July 1945. The U.S. government used its products, however, not to defeat Germany, which had already surrendered two months earlier, but to end the war with Japan.

Out of the closely knit, highly focused teamwork that built the atomic bomb came a group of physicists that included both seasoned senior scientists (such as Enrico Fermi and Hans Bethe) and young, innovative minds (such as Richard Feynman).

or in an office that have 30 to 100 times the memory and speed of the 1940s computers that filled rooms the size of a gym and cost millions of dollars. In the 1940s, too, all computers were programmed using hard wiring and they stored data in big magnetic core devices. Data was entered using punched cards. The first programming language was invented in 1956, enabling one computer to have many different functions, and by 1970, direct keyboard entry was possible and data could be stored on floppy disks. By the 1990s, fixed disks, or hard drives, with storage capacity ranging from 120 to 200 million bytes became common in desktop computers.

Today, few businesses—even one-person entrepreneurships—get along without the use of at least one computer. Millions of households have them. And for scientists, the personal computer and the larger, mid-range computer have become a great asset in building theories and exploring possibilities through the use of computer modeling. By building a model, they can play out complicated "what-if" scenarios for everything from planetary formation to predictions about the movement of tectonic plates or the future course of the greenhouse effect.

The most enormous impact of solid-state electronics technology was on computers, but solid-state devices, or microprocessors, are everywhere— in radios, TVs, microwave ovens, automobiles, video tape recorders, record players, audio cassette recorders and CD players, telephones, programmable thermostats, and on and on. And more applications turn up every day.

Richard Feynman earned fame early as one of the rising stars who fought the good fight in what was informally known as the "War Against Infinities"—the struggle to find a theory that united quantum theory with James Clerk Maxwell's highly successful 19th-century electromagnetic field theory.

By 1946, the two great revolutions of 20th-century physics—that is, quantum mechanics and relativity—had already produced a profound affect on our mental images of subatomic particles. Werner Heisenberg, with his "uncertainty principle," had established that an electron could not be thought of as having both a particular velocity and a location at exactly some point in space; instead we know only a probability of where it might be. Also, according to quantum rules, a phenomenon known as "virtual particles" can be created by borrowing the necessary energy—coming into existence for the barest sliver of a second and then, just as suddenly, disappearing. So there can be a real electron, whose exact position we can never know, surrounded by a swarm of transient virtual photons. The photons (messenger particles of light) let us know that the electron is there. They also ever-so-slightly modify its properties, and these modifications can be measured with careful, precise measurements and analyzed by patient, determined theoretical calculations. All of this enormously complicates the process of measuring the physical, observable electron in an experiment.

If you find this confusing, you have lots of very intelligent company. "[Electrons and light] behave in a way that is like nothing that you have ever seen before," Richard Feynman would tell his students a few decades later. "Your experience with things that you have seen before is incomplete. The behavior of things on a very tiny scale is simply different." Simplifications and metaphors don't really work, Feynman added. Atoms aren't really like solar systems or springs or clouds. Only one simplification really works, he said: "Electrons behave in this respect in exactly the same way as photons; they are both screwy, but in exactly the same way . . .

"I think I can safely say that nobody understands quantum mechanics. . . . I am going to tell you what nature behaves like. If you will simply admit that maybe she does behave like this, you will find her a delightful, entrancing thing. Do not keep saying to yourself, if you can possibly avoid it, 'But how can it be like that?' because you will get 'down the drain,' into a blind alley from which nobody yet has escaped. Nobody knows how it can be like that."

And then Feynman would go on to show how the experiments, the calculations, all the evidence point to the fact that indeed this tiny world that works like nothing we know does in fact work like that. In fact, Feynman's work, quantum electrodynamics (known as QED for short), tied together in a theoretically tidy package all the phenomena at work in light, radio, magnetism, and electricity. At the same time, independently and

Richard Feynman at Los Alamos (Courtesy Los Alamos National Laboratory)

Shin'ichiro Tomonaga (in center with raised hand) shared the 1965 Nobel prize for physics with Richard Feynman and Julian Seymour Schwinger for his work in quantum electrodynamics. (*AIP Emilio Segrè Visual Archives*)

separately, two other people were working on the same theories: New York-born Julian Seymour Schwinger and Japanese physicist Shin'ichiro Tomonaga.

Schwinger (1918–), a child prodigy, entered the College of the City of New York at the age of 14. By the age of 21 he had completed his Ph.D. at Columbia University and at 29 he was promoted to full professor at Harvard University. He was the youngest in the long history of the university to attain that rank. Tomonaga (1906–79), meanwhile, a classmate of Yukawa's at the University of Kyoto, had studied for a time with Heisenberg in Germany, returning to Japan to finish his doctorate in 1939. During World War II, cut off by the strife from American and European physicists, Tomonaga worked on what would eventually become QED while teaching at Tokyo University of Education, of which he became president after the war, in 1956.

In the late 1940s, though, some of the calculations for electrons, as they interacted with virtual particles popping in and out of existence, came up with a result of infinity for the mass of the electron—an obvious error that everyone recognized as absurd for this tiny particle. Feynman, Schwinger and Tomonaga worked out the behavior of electrons mathematically with

11

RICHARD FEYNMAN: WELL-ROUNDED GENIUS
(1918–88)

Feynman, teaching at Caltech (California Institute of Technology)

Richard Feynman's formulation of quantum electrodynamics in 1948, while he was still in his twenties, earned him a share in the 1965 Nobel prize for physics. Without question, he was one of the most brilliant, unconventional and influential physicists of modern times. His original mind, boundless interests, alternately outgoing and contemplative personality, and unstoppable sense of humor set him apart from the beginning. In general, scientists are cautious about using the word *genius,* but most will apply it to two 20th-century physicists: Albert Einstein and Richard Feynman.

Feynman always had an urge to know how things worked. He fixed radios, picked locks, learned to speak Portuguese and deciphered Mayan hieroglyphics. In high school, his teacher let him sit in the back and work problems in advanced calculus while the other kids worked conventional physics problems. He was smart. But he was also known for his nonstop bongo playing, and he once played an impromptu xylophone, made by filling water glasses, and kept it up all evening to the great amazement of visiting Danish physicist Niels Bohr. As fellow physicist C. P. Snow described him, Feynman was "a little bizarre . . . a showman . . . rather as though Groucho Marx was suddenly standing in for a great scientist."

After completing his studies at the Massachusetts Institute of Technology and at Princeton, where he received his Ph.D. in 1942 after publishing his dissertation on quantum mechanics, Feynman was commandeered for the atomic bomb project. He soon became a brash young group leader and gained exposure, at close hand, to some of the finest physicists in the world, including Enrico Fermi, Hans Bethe, Niels Bohr and his son, Aage, and J. Robert Oppenheimer.

After the war, while teaching at Cornell University, he worked on QED, the theory for which he shared the 1965 Nobel prize for physics. Considered to be one of the key architects of quantum theories, he invented a widely used method known as "Feynman diagrams," which gave physicists a way of visualizing particles and their collisions and a way of talking about them in a common language. Many fellow physicists consider that at least three of Feynman's later achievements are also worthy of a Nobel: a theory of superfluidity, the frictionless behavior of liquid helium; a theory of weak interactions; and a theory of partons that helped produce the modern understanding of quarks. Like Einstein, he was always ready to accept the challenge of nature's next riddle.

Feynman also founded the field of nanosystems, the building of devices on a miniature, even molecular, scale—resulting in a new and highly successful industry, molecular manufacturing, which builds complex structures with atom-by-atom control. Because nanotechnology does for manipulating matter what the computer did for manipulating data, the potential is enormous. Following the explosion just seconds after liftoff of the U.S. space shuttle *Challenger,* on January 28, 1986, Feynman played the part of caustic inquisitor in the presidential commission's investigation. He graphically pointed out, with a demonstration using ice water and an O-ring, that the rubber used in these critical seals became brittle and unable to hold a seal at low temperatures—and that that was what had caused the disaster. The morning of the *Challenger*'s launch was cold—icicles hung from the scaffolding—and as the shuttle headed skyward, fuel leaking through a faulty seal in one of the booster rocket joints ignited and caused the explosion, killing the seven people aboard. Though ill, Feynman was tireless in pursuing the series of shortcuts and poor decisions that had resulted in ignoring the physical limitations of a piece of rubber on which rested seven lives and a multi-million-dollar space program.

Feynman's autobiography, *Surely You're Joking, Mr. Feynman,* published in 1985, was a surprise runaway best-seller, and after his work on the *Challenger* commission, he followed up with another, *What Do You Care What Other People Think?* Although he was not able to complete the writing before his death, the book still provides a final glimpse inside a great and original mind.

Perhaps most important of all was Feynman's legacy as a superb lecturer and teacher—his unique approach to problem solving and life influenced many younger scientists at both Cornell and Caltech, where he later taught, and his published lectures have become classics.

new theoretical insight and far greater precision than anyone had done before, and they overcame the error. Their calculations could explain the electromagnetic interactions of electrons, positrons and photons with stunning accuracy. Could the same be done for neutrons and protons, held together in the nucleus by a powerful bond known as the strong force? Hopes were high.

PARADE OF PARTICLES

But the experiments did not go well. The strong force, it turned out, was much more complicated than anyone had imagined.

By 1941, Yukawa and the rest of the scientific community had realized that the meson found in 1936 by Carl Anderson was not the predicted carrier of the strong force. It was something else. Shortly before the Japanese attack on Pearl Harbor on December 7, 1941, Yukawa, in Kyoto, wrote dejectedly, "The mesotron theory," (as he called it) "is today at an impasse."

The war slowed down both communication among scientists and research, although three Italian physicists managed to conduct an experiment secretly in a Roman cellar that proved that Anderson's mesons hardly interacted at all with the nuclei of atoms. When they finally were able to announce their results after the war in 1947, the search was on, again, for Yukawa's meson.

It didn't take long. After the war, a British chemical company had begun producing photographic emulsions capable of revealing high-energy cosmic rays, and Cecil F. Powell in Bristol led a team that used these emulsions to trace the tracks of cosmic rays. A charged particle passing through the emulsion leaves a trail of ions that induce black grains to form after development. From the number and density of the grains, Powell and his colleagues could deduce some of the particle's characteristics, such as mass and energy. And when they looked at the cosmic-ray particle tracks, they found evidence for some that did interact, using the strong force, with atomic nuclei. What's more, their weight was close to the mass predicted by Yukawa—a little heavier than Anderson's meson. Using the Greek letters *pi* and *mu* to distinguish the two middle-weight particles, Powell called the new particle a pi-meson, soon contracted to *pion*, and Anderson's meson a mu-meson, which became *muon*. The year was 1947, about the same time that Feynman and the rest were firming up QED to explain and predict electron behavior, and excitement began to build that perhaps a similar breakthrough was at hand for particles in the atomic nucleus.

Not everyone was ecstatic, though. Powell's discovery meant that Anderson's muon was "extra," apparently unnecessary, according to all current theories. As Columbia University physicist Isidor Rabi put it quizzically, "Who ordered that?"

Instead of clearing up, the story of the nucleus steadily became even more confusing. After the pion, physicists began to discover whole families of particles related to it, or related to the proton. Each new particle discovered brought with it a realization that the virtual cloud swarming about the nucleus must be even more complex than previously thought, and the mathematical equations describing the interactions began to become hopelessly complicated. In 1947, two scientists at the University of Manchester spotted a particle in their cloud chamber that they dubbed the K-meson, or kaon, related to the pion. (A cloud chamber is a device in which the paths of charged subatomic particles are made visible as trails of droplets in a supersaturated vapor.) Two years later, Cecil Powell's team found tracks in their emulsions of a charged particle that broke down into three pions, and they called their new particle the tau-meson. Not until 1957 did anyone realize that the two were just different states—positive and negative versions—of the same particle, which finally was called a kaon. Cosmic ray physicists discovered yet another particle, apparently a neutral cousin of the positively charged proton, in the early 1950s, which they called a *lambda*.

In the midst of this chaos a new and powerful tool was about to make its mark. So far, most of the discoveries had been made by cosmic ray physicists tracing particle tracks in their cloud chambers. But to answer the questions now being raised, particle physicists needed more detailed data than the cloud chambers alone could produce. That's where particle accelerators came in. These powerful machines could deliver uniform, controlled beams

LASERS: QUANTUM PHYSICS AT WORK

One of the stunning outgrowths of quantum mechanics are lasers, which have only been around in a workable form since Theodore Harold Maiman developed the first laser in 1960. A *laser* (the word stands for *Light Amplification by Stimulated Emission of Radiation*) produces a narrow, accurate beam of very bright light—different from ordinary light because it is coherent. That is, all rays of a laser have exactly the same

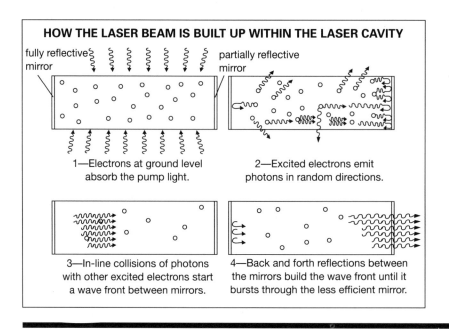

HOW THE LASER BEAM IS BUILT UP WITHIN THE LASER CAVITY

fully reflective mirror

partially reflective mirror

1—Electrons at ground level absorb the pump light.

2—Excited electrons emit photons in random directions.

3—In-line collisions of photons with other excited electrons start a wave front between mirrors.

4—Back and forth reflections between the mirrors build the wave front until it bursts through the less efficient mirror.

of high-energy particles—electrons, for example, or protons or pions. And by tracing the results when they smashed into each other, they could find out an enormous amount of accurate detail about their properties. In fact, physicists can do two basic types of experiments using an accelerator and a particle detector: scattering experiments and particle-producing experiments.

In scattering, experimenters look for clues about the nucleus by tracking how particles scatter: how many, in what directions and at what angles. The higher the energy from the accelerator, the better the structure comes into focus. Using this technique enabled scientists to find out about the compo-

wave length and their waves are perfectly in phase and parallel. Or, put another way, in a laser, the quanta (packets of which the light is composed) are organized end to end, stuck together, in a way, producing a continuous wave. That's why, vibrating together, they produce light of very high intensity.

Laser beams can either be visible light or they can be infrared rays, which are invisible, and they have numerous uses in communications, engineering, science and medicine. Visible light lasers are used in recording, CD Rom players, and fiber-optic communications, while an infrared laser can cut through substances ranging from metal to human tissue with extremely fine accuracy.

A laser beam is produced by exciting atoms in a medium that can absorb and release energy. A surprising variety of substances will work as a medium: a solid crystal such as a ruby, certain liquid dyes or a gas such as carbon dioxide.

Energy is introduced to excite, or pump, the atoms of the medium, raising them to a high-energy state. For example, electrons from an electric current introduced into a gas medium excite the atoms of gas. Finally, one or more atoms reach a higher energy and release a light ray (or photon). The light ray strikes another atom, which reaches a new level of energy and it, in turn, emits a light ray, and so on as the light builds up. (This process is known as stimulated emission.) Mirrors—one fully reflective and the other only partially reflective—reflect the light rays back and forth (in the part of the process known as amplification), so that more and more atoms release light.

Each new ray emitted by an atom vibrates in step with the ray that struck the atom. With all the beams in step, the light escapes the tube through the partially silvered mirror, and the energy is released as laser light.

sition of nuclei—how the protons and neutrons were arranged and how they jostled within the nucleus to maintain their arrangement. With still higher energy, experimenters could see even deeper, to see how the parts of the protons and neutrons fit together.

The second use of particle accelerators and detectors—the discovery of new particles—proved effective immediately. By 1949, scientists in Berkeley, California isolated the neutral pion using the big synchro-cy-clotron built there under the direction of Ernest O. Lawrence. It was the first new particle to be found at an accelerator instead of in cosmic-ray experiments.

SUPERCONDUCTORS

In 1911, a Dutch physicist named Heike Kamerlingh discovered that when mercury is cooled to very low temperatures—close to absolute zero (-273.15° C, -459.67° F, or 0° K)—electrical resistance vanishes. He had discovered a strange phenomenon that became known as superconductivity—one of the most important breakthroughs in modern experimental physics.

Not until 1957, though, did anyone come up with a theory explaining why this works. Three U.S. physicists, John Bardeen, Leon N. Cooper, and John Robert Schrieffer, proposed what is known as the BCS theory (named after them), stating that the superconducting current is carried by electrons linked together through lattice vibrations that cannot dissipate energy through scattering, the usual cause of electrical resistance in conductors.

Practically speaking, though, the fact didn't have much use at first, since the temperatures at which this phenomenon went into effect for various substances, known as the "critical temperature," were all so low. Superconductors were put to use in large, powerful magnets used in particle accelerators and magnetic-resonance imaging (MRI) machines used in medicine, but they are complicated and expensive because they have to be cooled with liquid helium.

Then in 1986 and 1987, two researchers at IBM, K. Alex Müller and Georg Bednorz, reported the development of materials that become superconducting at substantially higher temperatures than liquid helium, which boils at 4.2° K (-268.95° C or -452.11° F). These new materials, which are ceramics, have a critical temperature between 90° and 120° K, another big breakthrough, since these temperatures are above the boiling point of liquid nitrogen, which is much cheaper and easier to maintain than liquid helium. These high-temperature superconductors lose their superconductive properties, though, at high current levels, a problem that has hampered their use in commercial applications.

In the fall of 1955, researchers Emilio Segrè and Owen Chamberlain succeeded in spotting the negatively charged antiproton, the proton's antimatter twin. It had been almost 25 years since Carl Anderson had found the electron's twin, the positron, on August 2, 1932. Protons were accelerated at the University of California at Berkeley in the newly built Bevatron and were hurtled at a copper target at an energy of six billion electron volts. (This is the minimum level required to coax an antiproton into appearing from a flash of energy.)

Emilio Segrè
(University of California
Lawrence Radiation
Laboratory, Courtesy AIP
Emilio Segrè Visual Archives)

In the 1950s and 1960s, a huge swell of new subatomic particles flooded the scientific journals from the accelerators at Berkeley, Brookhaven (on Long Island, New York), Stanford, Fermilab (near Chicago) and CERN (in Geneva). The more particles physicists found, the more undiscovered particles they found evidence for, and often the particle around the next corner promised to be heavier and more resistant to discovery.

The higher the energy of the accelerator, the more likely that the physicists could break the next particle loose, as they moved deeper and deeper into the structure of the nucleus. Lawrence's 1949 synchro-cyclotron wielded a stream of particles at 100 MeV. (MeV means million electron volts, usually pronounced emm-ee-vees. An electron volt is a measure of the amount of energy required to move an electron across a potential difference of one volt.) Today, the Tevatron at Fermilab can bump its energy up to one TeV (a trillion electron volts).

ALBUQUERQUE ACADEMY
LIBRARY

19

TOOLS OF SCIENCE: THE PARTICLE ACCELERATOR

1929   British physicists John Cockcroft and Ernest Walton invent the first particle accelerator, a simple machine that enables them to bombard lithium and produce alpha particles (helium nuclei), artificially transforming lithium into helium. They are able to confirm the production of helium nuclei by observing their tracks in a cloud chamber.

1930   Ernest O. Lawrence at the University of California at Berkeley comes up with his idea for the cyclotron, which can accelerate protons or alpha particles in spiral paths between two circular poles of a large electromagnet. The first cyclotron is very small, able to fit in the palm of his hand. The spiral path makes powerful acceleration possible without high voltage or a superlong, straight path. Larger and larger versions follow throughout the 1930s at Berkeley's Radiation Laboratory (now known as the Lawrence Berkeley Laboratory). By 1939 they have a 60-inch-diameter cyclotron that can boost particles up to 10 MeV. Soon Lawrence begins planning a 184-inch cyclotron capable of reaching 100 MeV, an unheard-of energy, and a new "synchro-cyclotron" version follows, capable of hundreds of MeV.

1952   A consortium of eastern universities builds the first "synchrotron," the Brookhaven National Laboratory's "Cosmotron," on Long Island, New York. The big machine is capable of delivering more than one billion electron volts (1,000 MeV or 1 GeV, for giga electron volts, meaning a billion), later as much as three billion. Instead of having huge fixed magnets through which particles are made to orbit, the synchrotron's magnets are smaller and moveable—synchronized to move into position as the particles' orbits intersect them. This means that while

---

Experimenters also developed a vast supply of variations to get at particular pieces of information—different kinds of detectors, different kinds and energies of "bullets," to find out a particle's lifetime, decay modes and so on. (All the new particles were very unstable and broke down quickly into other particles.) The data was overwhelming.

Particle physics seemed to be headed for total disorder and confusion.

cyclotrons are limited by the size of magnets one can manufacture, the size of the synchrotron is only limited by the consortium's finances and the size of the land parcel it's built on.

1954   The Conseil Européen pour la Recherche Nucléaire (European Council for Nuclear Research), now called the European Organization for Nuclear Research but still commonly known as CERN, is founded in Geneva, Switzerland.

1959   CERN begins operation of its 25-GeV accelerator.

1965   The SLAC (Stanford Linear Accelerator Center) particle accelerator goes into operation in California.

1969   Robert Wilson founds the Fermi National Accelerator Laboratory, soon known as Fermilab, in Batavia, Illinois, near Chicago.

1972   The large accelerator at Fermilab begins operations at 200 GeV (giga electron volts, or 200 billion electron volts). Later that year it reaches 400 GeV.

1976   CERN begins operation of the Super Proton Synchrotron (SPS), a 4-mile ring originally delivering protons at 300 GeV, eventually reaching 450 GeV.

1984   The particle accelerator at Fermilab reaches 800 GeV.

1985   The 900-GeV Tevatron particle accelerator at Fermilab begins operation; later it is boosted to 1 TeV (1 trillion electron volts).

1985   A team headed by Carlo Rubbia at CERN discovers the charged W particles and the neutral Z particle, predicted by the electroweak theory.

THE REALM OF QUARKS

*P*athways through the chaotic subatomic particle jungle have not been easy to find. But to one multitalented physicist goes much of the credit for blazing those pathways—ingeniously sensing the topography and insightfully mapping it out, while marking the complicated trail with whimsical names and literary allusions.

Murray Gell-Mann, whose father had left Austria to settle in New York, was born in New York City in 1929. He entered Yale University on his 15th birthday, and that fact alone tells a lot. At age 21, he had earned a Ph.D. from the Massachusetts Institute of Technology. And, after further studies under Enrico Fermi in Chicago, Gell-Mann attained the rank of full professor at the California Institute of Technology (Caltech) by the time he was 27. Having a razor-sharp intellect, highly eclectic interests and a gift for languages (fluent in many, including Swahili), he's been known to savor reminding his more "ordinary" colleagues of the I.Q. gap.

By the time of his arrival at Caltech, Gell-Mann had already plunged deep into the jungle of particle physics. In addition to Chadwick's neutron, Dirac's positron and Pauli's neutrino, Yukawa had postulated the meson— of which too many had turned up: not only Anderson's mu-meson, later renamed "muon" because it turned out not to be a meson; but Powell's pi-meson (pion), which was Yukawa's carrier of the strong force. By the 1950s, there were also K-mesons, heavier yet, which had about half the mass of a proton. Before long, particles even heavier than protons—called hyperons—began to turn up.

And it was these K-mesons, or kaons, and hyperons that especially intrigued Gell-Mann in the early 1950s. These particles, he reasoned, were created by strong interactions and should, by rights, also be broken down by them. But they were not. Instead they were broken down by the weak interactions (the interactions of which radioactive emissions are the evidence).

When Marie and Pierre Curie began studying radioactivity back in the 1890s, they meticulously measured the results of the mysterious "beta ray" emission (release of electrons from the nucleus) that few people (aside from colleagues such as Henri Becquerel) had ever even heard of. By the 1950s, though, a great deal was known about radioactivity and the force that governed it, called the weak interaction. Measured at about 1,000 times weaker than the well-known electromagnetic interaction and weaker yet than the strong interaction that holds nuclear particles together, the weak interaction was a reasonably well understood phenomenon—or at least most physicists thought so.

Well understood, that is, except for the fact that, by all rights, the definitely weaker and ever-so-slightly slower weak interaction should not prevail over the faster strong interaction. Kaons should have decayed via the strong interaction, according to what was known, but they didn't. They only decayed via the weak interaction. These facts seemed really strange to particle physicists. And, as a result, they started calling kaons and hyperons "strange particles."

Murray Gell-Mann (California Institute of Technology)

23

LOOKING AT STRANGENESS

So in the early 1950s Gell-Mann became immersed in questions of strangeness. (At the same time, Japanese physicists T. Nakano and Kasuhiko Nishijima were working along the same lines independently and came to similar conclusions.) First, he began to think in terms of groups of sub-particles, instead of thinking of them individually. If you looked at the characteristics of protons and neutrons, for example, they were strikingly similar in every way—except that one was positively charged and the other was neutral. He found that, if you ignored the charge of the subparticles, most sub-particles in the nucleus seemed to fall into groups of two or three.

So Gell-Mann divided the known particles into groups based on all characteristics except charge. Then, based on the total charge of all members of each group, he assigned each group a charge center. The neutron-proton group, for example, had a charge center of +1/2 (since the group's total charge was +1 and there were 2 members). But for kaons and hyperons, strangely, the charge center wasn't really in the middle the way it was in the other groups—it was off-center. Gell-Mann found that he could measure just how much off-center they were and he assigned a number to the degree they were off-center—a number he called their "strangeness number." The strangeness number for protons and neutrons was 0, since they aren't off-center at all. But he found that some particles had a strangeness number of +1, -1 or even -2.

What's more, Gell-Mann noticed a pattern in all particle interactions: The total strangeness number of all particles involved in any interaction is always conserved; that is, it's the same before and after the interaction. Physicists liked this because it illustrated that a kind of symmetry was at work (which nature often displays) that could be described in quantitative terms (which physicists always like), and they could use it to explain the unexpectedly long life of the strange particles. Both Gell-Mann and the Nakano-Nishijima team published their ideas along these lines in 1953.

Some mysteries about the weak interaction remained unexplained, though—mysteries that came up one afternoon in 1956 over lunch as Chen Ning (Frank) Yang and Tsung Dao Lee chatted at the White Rose Cafe in New York City. As these two longtime friends and colleagues talked, they began to have a sneaking suspicion about the weak force that no one had ever thought of before.

A LEFT-HANDED WORLD

Frank Yang, who was born in Hofei, China in 1922, had traveled to the United States at the age of 23 to study with Enrico Fermi. But he arrived at

New York's Columbia University only to find that Fermi had left for the University of Chicago. Unflappable, Yang set off for Chicago, where he did study under Fermi, earning his Ph.D. in 1948. That's also where he met up with Tsung Dao Lee, whom he had first encountered in China. By 1956 he had already established his reputation when, in 1954 with Robert Mills, he helped set the stage for quantum field theory with the establishment of what are sometimes called Yang-Mills gauge invariant fields.

Lee was born in Shanghai in 1926 and had traveled to the United States in 1944 to study in a graduate program—even though he had not yet completed an undergraduate degree. The University of Chicago was the only school that would let him enroll; the university's acceptance was lucky for Lee, since some of the best minds in physics were there at the time. He was well prepared to take advantage of that fact and obtained his Ph.D. in 1950, working under hydrogen bomb architect Edward Teller.

Lee and Yang crossed paths again for a time at the Institute for Advanced Study in Princeton, New Jersey, where Yang stayed to become professor of physics in 1955, while Lee accepted a position at Columbia University in

Chen Ning (Frank) Yang (left) and Tsung Dao Lee (Alan Richards, Courtesy AIP Emilio Segrè Visual Archives)

New York City in 1953. For New Jerseyites, New York is always just a train ride away, and the two generally got together once a week to compare notes.

The topic of conversation on that particular afternoon at the White Rose Cafe was the "strange particles" called K-mesons, which seemed to break down in two different ways—one, a sort of right-handed way and the other, a sort of left-handed way.

This ordinarily shouldn't happen—and it didn't happen with other particles. The way that the K-mesons broke down seemed to defy an important principle of physics: the law of conservation of parity. Like the laws of conservation of energy and conservation of matter, conservation of parity was a guiding principle that always before had seemed to predict nature consistently.

Imagine yourself standing in front of a mirror. What in reality is on your right is mirrored on your left, and your left side appears in the mirror to be on the right side of the mirrored image. If your hair is parted on the right, in the mirror, it appears to be parted on your left. Now imagine reversing the rest of the image, top to bottom and front to back. The law of conservation of parity says that if you take a system and switch everything around in this way, the system will still exhibit exactly the same behavior.

Parity has two possible values: odd and even. The law of conservation of parity says that if you start out with odd parity before a reaction or change, you should end up with odd parity afterward. That is, when particles interact to form new particles, the parity should come out the same on both sides of the equation.

The problem with the K-mesons was that when they broke down, sometimes they broke down into two pi-mesons, both with odd parity (which combined to give even parity). Other times they broke down into three pi-mesons (which combined to give odd parity). It was as if you looked in the mirror and sometimes your right hand was reflected on your right and sometimes on your left. The two sides of the equation were supposed to mirror each other perfectly, but they didn't always. Physicists tried to fix this by saying that there were two different types of kaons, one with odd parity and the other with even parity. But Yang and Lee thought that might not be the right solution. These mesons were exactly the same in every other way. Maybe there was something else at play.

Was it possible, Lee and Yang asked each other, that conservation of parity did not apply to these "strange particles"? Maybe there actually was just one K-meson, not two. Maybe the reason parity of conservation appeared to be broken was that the principle did not apply to the weak interaction. They realized that no one had ever checked out this possibility, and they began to think about the kinds of situations that might test the premise. And so began what became known as "the downfall of parity"—not a total demise, but a downfall with respect to one area, the realm of the weak interaction.

Lee and Yang began sketching out a paper, which they shortly published, on "The Question of Parity Conservation in the Weak Force." In it they reviewed a number of reactions and examined the experimental implications of the possibility that parity (that is, mirror symmetry) was not respected by the weak force. How could you test this idea? They thought if you could examine the directions in which an emerging electron is ejected from a spinning nucleus in beta decay (the realm of the weak interaction), and could see, for example, that the electron favored one direction over the other, then that would be a tip-off.

The original idea—the theory—had sprung from the minds of Lee and Yang in collaboration. But a theory is only worth something in science if it stands up to experiment. If it does, it may open up a large new area of research, generating new and provocative questions and putting to bed many old ones.

Immediately, Chien-shiung Wu, the experimental counterpoint to Lee and Yang, swung into action. Professor of physics at Columbia and a colleague of Lee's, Wu was a well-respected, iron-willed experimentalist,

Chien-shiung Wu, "Madame Wu," whose experiments showed that Lee and Yang's ideas about the conservation of parity were right, shown here in 1958 performing an experiment on beta particles (Columbia University)

whose specialty was radioactive decay. Known for being tough on her students and demanding, she was equally demanding of her own work and exceedingly energetic. In this case, Wu—or Madame Wu, as she was almost universally known (even though Wu was her own name; her husband's name was Yuan)—ran experiments that were timely, intricate and clean. Wu decided to use cobalt 60, which decays into a nickel nucleus, a neutrino and a positron. What Wu wanted to "watch" with her apparatus was the spin of the positron as it sped away from the nucleus—but she had to make sure that the nuclei of her cobalt 60 samples were all spinning in the same direction so that the spin of the nuclei would not affect the spin of the emitted particles. To do this, Wu planned a complicated experiment that involved using cryogenics at the U.S. Bureau of Standards in Washington, D.C. to drop the cobalt to a very low temperature, just above absolute zero.

By early 1957, she had begun to get stunning results. "Wu telephoned," T. D. Lee told his colleagues over the first weekly departmental lunch of the new year, "and said her preliminary data indicated a huge effect!" Before long, Wu's results were in. Parity did not hold for the weak force. And by the end of the year Lee and Yang had won the Nobel prize for their insights.

Many physicists were not happy, however. The subatomic world, unlike the disorderly everyday world, had always seemed to have a wonderful elegance, a predictability—a symmetry—about it. Now symmetry seemed to be a sometime thing.

"I cannot believe God is a weak left-hander," Wolfgang Pauli quipped in disgruntlement. (Not that he thought being left-handed was undesirable—just that he had always seen nature as evenhanded.) He voiced an unease that many physicists felt. Many others began to wonder about the consistency of other laws of conservation. If parity isn't consistent, then maybe the other conservation laws are not always consistent. Maybe symmetry cannot be counted on as a principle anywhere. Lee, Yang and Wu had raised many questions, many of which remain unanswered still. But for scientists, who thrive on trying to find answers to unanswered questions, it's the sign of good science, not only to answer questions and make the pieces fit, but to raise new ones.

THE CORKER: THE QUARK

Meanwhile, in California, Murray Gell-Mann had been busy as well. A great theorist has a special talent for synthesizing and spotting patterns amid confusion, and this is exactly what Gell-Mann did. Several things needed sorting out and explanation, including the mind-boggling number of particles (why so many?) and the apparent groupings into families (what mechanism or principle caused them?).

Based in part on the work of Yang, Lee and Wu, Gell-Mann came up with some ideas, a system of classification, which he published in a long series of papers in the early 1960s. He called his system the "Eightfold Way," a term he borrowed from an aphorism attributed to Buddha in Chinese literature. (Not, as some enthusiasts have thought, that Gell-Mann meant to imply that physics had become mystical or even philosophical—he just needed a name for a concept that was so new to the world of language that its very name must be invented. Most of the letters of the Greek alphabet were already taken up with the task of naming particles. So he borrowed from one of his many interests.)

Gell-Mann's line of reasoning went like this: As he had already noted, many subatomic particles—including mesons, protons and neutrons—occurred in families, groups of two or three. There were three pions (pi-mesons), two pairs of kaons (another way of saying K-meson), a pair of protons (proton and antiproton), and so on. These were very close-knit families, with strong resemblances. In fact the members of a family were more alike than they were different. The only differences in each case were charge and mass. And the difference in mass was so small (a few MeV) that they could be caused simply by the difference in charge. In other words, these particles were in all likelihood identical, since the difference in mass could be caused by the difference in charge alone. So, said Gell-Mann, what if you thought of each of these families as a single particle with another characteristic—"multiplicity"? It was a new and productive way of looking at the variety of particles to be found in an atomic nucleus.

Secondly, he noticed that the strong force pays no attention at all to electric charge. It has equal effect whether a particle is neutral, negative or positive. A proton is held as strongly as an antiproton. The strong force makes no distinction among the neutral pion, its positive sister or its negative brother. They are like three sides of an equilateral triangle.

Gell-Mann thought there was a connection between the strangeness found in the strange particles and their multiplicity. Unlike pions, which came in triplets, the strange kaons seemed to come in sets of two pairs. He was sure that some deeper symmetry, not yet discerned, was at play here; these were not just coincidences.

At the time, in late 1960, mathematicians had recently rediscovered the work of a Norwegian mathematician named Sophus Lie, who had explored some areas of an abstract formalism called "group theory." Gell-Mann recognized a particular Lie group—SU(3), or Special Unitary group in 3 dimensions—that seemed to work for both the mesons and baryons. (This same stroke of insight, by the way, came to Yuval Ne'eman at the Imperial College in London at about the same time.) Using the group as a sort of template, Gell-Mann arranged the mesons and baryons according to their charge and strangeness. But, while there were eight baryons, which fit perfectly into the pattern, there were only seven mesons.

So, based on the characteristics that an eighth meson should have if it fit the pattern, Gell-Mann predicted its existence, in much the same way that Dmitry Mendeleyev predicted the existence of several as-yet undiscovered elements when he drew up the periodic table of the elements in 1869. In particular, Gell-Mann predicted a particle he called an "omega-minus" particle, and he was right. Just such a particle was found in 1964—and observed again many thousands of times thereafter. Its antiparticle, anti-omega-minus, or "omega-plus," turned up in 1971.

And so, the Eightfold Way was born, and order, or at least greater order than before, was brought to the particle jungle.

But Gell-Mann had more up his sleeve. Even with the new order of the Eightfold Way, he thought there must be something more, some deeper and simpler order. There must be some particle more fundamental than anyone had recognized before. What physicists were doing, Gell-Mann recognized, was like looking at molecules of a substance and trying to understand their complexities without realizing they were composed of atoms (which is exactly what chemists did for centuries before Dalton). Baryons (neutrons and protons) were composed of something smaller—but what?

The answer began to emerge over lunch on Monday, March 25, 1963, at the Columbia University Faculty Club in New York City. (Apparently physicists do a lot of thinking while they eat!) Gell-Mann was visiting Columbia to give a series of well-attended lectures on the Eightfold Way and other issues, and some of the theorists of the host university, including Robert Serber, took him to lunch. Serber was a quiet man who had been at Berkeley with Robert Oppenheimer and had worked with him at Los Alamos. Generally, he preferred to work in the background, but on this day he had a question: What about a fundamental threefold group of particles, a triplet?

"That would be a funny quirk!" Gell-Mann replied off-hand. "A terrible idea," added T. D. Lee, who was also there. Then Gell-Mann began scribbling on a napkin: For a triplet to work, the particles would have to have fractional charges, a phenomenon never observed in nature and practically unimaginable. The particles would have to be +2/3, -1/3, -1/3.

But later he began to think more about it. As long as a particle didn't appear in nature with its fractional charge, maybe the idea wasn't so outlandish. If a truly fundamental nuclear particle, a fundamental hadron, were unobservable, incapable of coming out of the baryons and mesons to be seen individually, if it were forever trapped within the physical protons, neutrons, pions and so on that physicists had found in nature, then maybe it would work. He mentioned the idea in his next lecture. Back at Caltech, he worked further on the idea, and he mentioned it on the phone while talking to his old thesis adviser, Victor Weisskopf, who was then director at CERN, the particle accelerator in Geneva, Switzerland. Maybe baryons and mesons were made up of fractionally charged particles, Gell-Mann suggested. Weisskopf wasn't impressed. "Please, Murray, let's be serious," came his immediate reply. "This is an international call."

But Gell-Mann was serious. In 1964 he proposed the existence of an outlandish group of particles that carried fractional electric charges. Once again showing his flare for eccentric names, he named them *quarks*, based on a phrase from *Finnegans Wake* by James Joyce: "Three quarks for Muster Mark!" The one with a 2/3 positive charge he called an *up* quark. The other two he dubbed *down* and *strange*. A proton is composed of two up quarks and one down quark, having a total charge of +1. A neutron is composed of two down quarks and one up quark, resulting in a neutral charge. In the last sentence of his two-page article introducing the idea, he thanked Robert Serber for stimulating these ideas.

At the same time that Gell-Mann came out with his quarks, a young physicist named George Zweig was also working along similar lines. Zweig was an experimental physicist, working at CERN at the time, and he thought of these particles, which he called *aces*, as real, concrete particles, not abstract constructs, as Gell-Mann had. Because Zweig was younger and less well known, he didn't succeed in getting his revolutionary idea published (even Gell-Mann had chosen to submit to a lesser-known journal to avoid being turned down). But once word got around that Zweig had written an in-house paper at CERN on the subject, Gell-Mann always made sure Zweig got credit, although he teased Zweig about his "concrete block model."

Eventually, physicists came to the conclusion that, if there was a strange quark, as Gell-Mann had suggested, it must be one of a pair. So they began looking for what they called a "charmed quark," the companion to the strange quark. Amazingly enough, evidence for this idea came from not one, but two investigators at separate institutions: Samuel Chao Chung Ting at Brookhaven National Laboratory and Burton C. Richter at SLAC. Considering the fact that particle experiments usually require months, sometimes years, of planning and require the involvement of large numbers of scientists to carry them out, the possibility of two major facilities working on the same project without either one knowing it is very low. Yet, on the very day in November 1974 that Ting turned up at SLAC, planning to announce the birth of what he called the J particle, Richter announced that he and his team had found what he called a psi meson. Ting was astounded and, of course, not at all happy. They compared notes and found that the two teams, completely independently, had discovered the same particle, which they ended up calling the J/psi meson. Before long, investigators realized that, because of the properties possessed by the J/psi meson, it could not exist without the existence of the charmed quark. So Ting and Richter, independently, had not only found a new particle, but they had found evidence for the existence of a quark with charm. The Nobel prize for physics in 1976 was awarded to both Ting and Richter for their discoveries.

Flavor and Color

While Gell-Mann had sought to bring order to the chaotic jungle of subatomic particles, though, the number of new particles continued to increase. But they also continued to fit into the basic structure he had outlined.

A new view of the truly fundamental, structureless, unsplittable particle emerged: There were two basic kinds, quarks and leptons. However, there were three types (known as "flavors") of each. (The name *flavors*, of course, is another example of physicists' sense of humor and has nothing to do with taste.)

The three flavors of leptons were electrons (which scientists had known about for a while), muons (or mu-mesons) and tauons (or tau-mesons). Each flavor of lepton has four members; for example, the electron, the neutrino, the antielectron and the antineutrino.

Quarks were a little more complicated. Pairs of quarks became thought of as "flavors"—a concept that parallels flavors of leptons. So each flavor of quark has four members. For example, the up/down flavor includes the up quark, the down quark, the anti-up quark and the anti-down quark. If you think of the pairs of quarks as parallel to each of the three flavors of leptons, then the three flavors of quarks are up/down quarks, strange/charmed quarks, and, to fill out the scheme, a new pair: top/bottom quarks.

One of the key differences between leptons and quarks is that quarks feel the strong force and leptons do not. Also, leptons have integral (whole number) charges or none at all and do not combine. Quarks have fractional charges and apparently exist only in combination.

Now there was still one more big question about quarks in the 1970s. If quarks could never be found separated from the tight combinations they form, then what was binding them so tightly?

One powerful idea that virtually all physicists subscribe to is that each different flavor of quark comes in three varieties of a property that leptons do not possess. This property is called "color" (named by Gell-Mann), and the three varieties are called red, blue and green. These names are merely figurative; quarks don't literally have color as we know it.

But when quarks combine in groups of three, they always combine one each: red, blue and green. It's just a way to talk in threes. The red, blue and green cancel each other out to produce colorlessness (as the three primary colors do on a color wheel when you spin it).

Of course, quarks also combine in pairs to form mesons, and in that case, for example, a red quark will combine with an antired quark. The red and antired cancel each other out and you get a colorless result.

So color succeeded in explaining how quarks could sometimes combine in twos, to form mesons, and in threes, to form baryons. The study of this process—the way quarks combine in colors to produce colorlessness—is known as quantum chromodynamics (QCD). (*Chromo* comes from the Greek word meaning "color.") The QCD theory shows that color neutrality can be achieved by various combinations of quarks and antiquarks.

But how is color transported? What are the messenger particles that play the same part for the color-force that photons play for the electromagnetic force? Physicists came up with a particle called *gluons* ("glue-ons") which carry two versions of color: a color (red, green or blue) and its anticolor. When these gluons are emitted or absorbed by quarks, they change the quark color. The constant shifting back and forth of these gluons serves to hold (or "glue") quarks together with powerful force— made all the more powerful by the fact that as quarks move away from each other the force *increases*, decreasing when they come close together (exactly the opposite of the electromagnetic force). Try placing your fingers inside a rubber band. As you pull your fingers apart, the force between your fingers increases. The lines of color force act like the strands of rubber in a rubber band. As you move your fingers back together, the tension decreases—and it's the same way with the strong nuclear force carried by gluons.

SUBATOMIC PARTICLES

Hadrons (found inside the nucleus):
Baryons (protons, neutrons, hyperons)
Antibaryons (antiprotons, etc.)
Mesons (pions, kaons, etc.)

Quarks
(probably 6 types: up, down, strange, charm, bottom, top)
Antiquarks

Leptons (found outside the nucleus):
Electrons
Neutrinos
Muons

This complicated model of the atom—its many parts and how they are held together as described (very briefly) in the last two chapters—has become widely accepted among physicists. In addition to all the particles—hadrons and leptons and their sub-categories—four forces come into play in the atom: the strong force, which holds the nucleus together; the weak force, the force behind radioactivity; the electromagnetic force, which governs electric charge; and gravity, which operates only at long distances and has negligible effect on the interior of the atom.

Within the atom, messenger particles operate to convey the strong, weak and electromagnetic force. More recently, physicists have come to call these messengers "gauge bosons:"the photon for electromagnetism; W^-, W^+ and Z^0 (as predicted recently by the electroweak theory) for the weak force; and eight gluons for the strong force. Now that quarks are with us, the old idea that protons and neutrons are bound together by a pi-meson turns out to be incomplete. Protons and neutrons, as we've seen, are made up of quarks, and the messengers between these quarks are called gluons—operating at a much more fundamental level than was recognized earlier in the century.

Abdus Salam at a conference in Rochester, New York (AIP Emilio Segrè Visual Archives, Marshak Collection)

But there are problems with this model; not everyone is happy with it. For one thing it is incomplete. Not all the particles described by it have been detected, even indirectly. The top quark, for example, still was not experimentally confirmed as of mid-1994. No one had yet found the tau neutrino, either. Do neutrinos have a rest mass? No one had been able to tell. Other inconsistencies have also required some patchwork (always a bad sign for a theory or model).

Secondly, this model is extremely complicated—usually another bad sign. Physicists tend to think that simplicity, not complexity, is the rule, and wherever they have explored this idea thoroughly, nature has proven to prefer simplicity. Why, at the very heart of everything, would things get so complex? Most physicists think that, somehow, we have not quite arrived at the true, unified and simpler view of the world.

GRANDLY UNIFYING

Several people have tried, though, to come up with this simpler view of the world. Einstein spent the last years of his life trying to come up with a grand unified field theory (often referred to as a GUT) that would tie all the forces of nature together, but he never succeeded.

Sheldon Glashow   (Harvard University News Office; photo by Jane Reed, 1989)

In the 1960s American Steven Weinberg and Pakistani-British physicist Abdus Salam independently developed a theory of electroweak interaction that combined theories about electromagnetic interaction and the weak interaction, and Sheldon Glashow, who had attended the Bronx High School of Science with Weinberg, refined it (1968). Their theory provided a mathematical roof over the two interactions, and it was hailed as the first successful step toward the grand unified theory that Einstein had sought. And, while it wasn't completely proven right away, enough experimental support for it came in that the three men received the Nobel prize for physics in 1979. Then in 1983, using the accelerator at the CERN laboratory in Geneva, Carlo Rubbia and Simon van der Meer succeeded in discovering the W-particles (W^-, W^+ and Z^0) predicted by the electroweak theory. It was the final confirmation the theory needed.

Since that time attempts have accelerated to arrive at a unified theory of all four forces, and the search for GUTs has opened up many avenues for

Steven Weinberg at a meeting of physicists in Chicago, February 8, 1977 (AIP Emilio Segrè Visual Archives)

exploring what may have happened at the time the universe began—in the first few seconds of its existence and in the time directly following. The breakthroughs in particle physics throughout the last five decades have produced a highly fruitful cross-fertilization between physicists and cosmologists, each group excited about and contributing to cutting-edge theory and experiment in the other's field, as we'll see in the next chapter. But first, let's explore some of the ways in which astronomers and cosmologists have found out more about the universe and its objects in the second half of the 20th century.

STARS, GALAXIES, THE UNIVERSE, AND HOW IT ALL BEGAN

Since the beginning of human existence, as far as we know, people have looked with fascination at the far-off objects in the nighttime sky—watched them, learned their habits, plotted out their patterns and assigned meaning to their arrangements. If you lie on the grass in a mountaintop meadow on a dark, clear night, as many earlier people must have done, you can see that the sky holds countless complexities. With only a few instruments to help, the ancient Babylonians and Egyptians made many sophisticated observations. But once Galileo added the use of a telescope to the job of watching stars and planets in the 17th century, information about the world beyond our world began to multiply astoundingly. The planet Jupiter had moons, Galileo discovered, and the planet Saturn, it turned out, had rings. As telescopes improved, astronomers began to spot new structures and discovered, moreover, that there were several more planets than anyone had ever thought. By the 19th century they expanded their bag of tricks with photography and spectrography (used to study the distribution of energy emitted by radiant sources, breaking light up into its component colors, arranged in order of wavelengths).

But the 20th century, in a continuing interplay between theory and experiment and a marriage of astronomy and physics, has completely transformed our understanding of the vast realms of the universe. In the first half of the century, Einstein's theory of relativity taught us that we live in a space-time continuum that is shaped by the mass of objects. And the series of advances in quantum theory and nuclear physics in the first few decades of the century set the stage for extraordinary new ideas about how the universe began and its early history. Advances in instruments and methods of observing the skies, meanwhile, brought on discoveries of new types of objects never before dreamed of. More than ever before, the study of astronomy, astrophysics and cosmology (the study of the origins and structure of the universe) is reserved for those who are stout of heart and daring of mind. The statement that "there are more things in heaven and earth, Horatio, than are dreamt of in your philosophy"—as Shakespeare's Hamlet admonishes his friend—could easily be the motto of astronomers, as well as physicists, in the second half of the 20th century.

MORE THINGS THAN ARE DREAMT OF . . .

In the early 1950s, night after night, astronomer Allan Sandage rode an elevator to a spot high up under the dome of the Hale Observatory on Palomar Mountain, near Pasadena, California, to sit in a contraption known as the prime focus cage, the observer's spot at the big, 200-inch telescope. The cold of the mountain air numbing his fingers and toes, Sandage prized the solitude, the intensity of the work, and the feeling of being in the cockpit of a time machine. It was his nightly date with the stars and he never missed it.

Sandage and his colleagues had at their disposal the best equipment ever built for optical astronomy. The 200-inch (5-meter) Hale telescope had just been completed in 1948. (It is still the second largest reflector in the Western Hemisphere; the largest in the world is the 400-inch reflector known as the Keck, built in 1991 on Mauna Kea, a volcanic mountain near Hilo, on the island of Hawaii.) And Sandage had learned his trade on the excellent 100-inch telescope at nearby Mount Wilson, under the tutelage of Milton Humason. Hired on as an assistant to Edwin P. Hubble, the grand master of galaxy measurement, Sandage had embarked on a project that promised to last a lifetime. Hubble, who had succeeded in measuring distances to nearby galaxies, had undertaken a long-term observation plan to measure distances to more distant galaxies and, ultimately, to measure the size of the universe. His discovery, called Hubble's Law, states that the farther away a galaxy is, the greater the degree to which its light is shifted toward the red end of the spectrum—that is, the faster it is speeding away from us. This "red shift" of light is actually a Doppler shift, like the trailing whistle of a passing train, or the horn of a speeding car, as it races away; in this case a shift in light is produced by the galaxies' motion as they rush away from one another while the universe expands.

Sandage's job was to photograph galaxies and search for variable stars within them so that he could measure the distances between them. When Hubble died in 1953, Sandage had only just begun. But Sandage inherited Hubble's time on the 200-inch and all his charts and records, and

NEW WAYS OF SEEING

As we gaze up from Earth, even from a telescope perched high on a mountain and far from city lights, we peer through a haze of atmosphere that distorts and clouds the view. Many objects can't be seen clearly, and some kinds of radiation outside the visible light range cannot be detected at all. But with the birth of space rocketry in 1957, for the first time in history it became possible to observe from outside that atmospheric shell.

In July 1962, Riccardo Giacconi and his colleagues put an X ray detector aboard an exploratory rocket to see if they could find evidence of fluorescence on the Moon. In the process the little sounding rocket launched from White Sands, New Mexico made the first discovery of a cosmic X ray source, Scorpius X-1. (The name means that it is the first X ray source discovered in the constellation Scorpius, a constellation in the Southern Hemisphere, located partly in the Milky Way.) Finding an object in a nonvisible range of the electromagnetic spectrum like X rays is a lot like hearing someone knocking at the door, but not seeing anyone there. Until you open the door, you don't know who's knocking. In 1967, astronomers matched up Scorpius X-1 with a visible object, a variable star known as V 818 Sco. A second source, Taurus X-1, was discovered in 1963 and soon turned out to be the Crab Nebula, a turbulent cloud of expanding gas and dust, left over from the supernova observed by Chinese and Japanese astronomers in 1054. A flurry of rocket and balloon explorations followed these discoveries, and by 1970, astronomers had found 25 to 30 sources spread throughout our galaxy. By December 1970, the first X-ray satellite, *Uhuru,* was launched; it discovered a large number of new sources, most of which turned out to be binary systems (composed of two companion stars).

In 1983, the United States's NASA cooperated with aerospace programs in the Netherlands and the United Kingdom to launch the Infrared Astronomical Satellite (IRAS), which surveyed all but 2% of the entire sky for sources in the infrared wavelength range of the electromagnetic spectrum. IRAS, which contained optics cooled by liquid helium, kept up its survey for nearly 11 months, until it ran out of helium. By the time the data was analyzed and organized, the catalogue of IRAS observations published in 1984 contained about 300,000 sources. And the discoveries were extensive, including a dust shell around the star Vega (possibly an early planetary system), five new comets, and extensive information about the kinds of objects that emit infrared radiation.

CHART OF THE ELECTROMAGNETIC SPECTRUM

Wavelength (meters)	Type	Colors
10,000,000		
1,000,000	Ultra-Low Waves	
100,000		
10,000	Government Communications	
1,000	AM Radio	
100		
10	Police & Fire	
1	TV & FM Radio	
.1	Radar,	
.01	Microwave Communications	
.001		
.0001	Infrared	Red
.00001		Orange
.000001		Yellow
.0000001	Visible Light	Green
.00000001	Ultraviolet Light	Blue
.000000001		Violet
.0000000001		
.00000000001	X rays	
.000000000001		
.0000000000001	Cosmic and Gamma Rays	
.00000000000001		

Wavelengths are expressed in meters.
This chart shows the extreme range of
recorded electromagnetic wavelengths.

This chart shows the full range of the electromagnetic spectrum, with figures on the left indicating the wave lengths (distance from the crest of one wave to the crest of the next). The light we see—including all the colors of the rainbow—represents only one small portion of the electromagnetic spectrum, with wavelengths of just a few ten-millionths of a meter. Ultraviolet rays, X rays and gamma rays have even shorter wavelengths than visible light, while infrared waves are longer and radio waves are much, much longer, up to a meter and more. In the last half of the 20th century astronomers have developed ways to observe in the microwave, infrared, X ray and gamma ray portions of the spectrum in addition to visible light.

The U.S. fleet of space shuttles, first launched in 1981, has provided a way to place several complex astronomical observatories into orbit, including the Hubble Space Telescope in 1990 and the Gamma Ray Observatory (GRO) in 1991.

The Edwin P. Hubble Space Telescope was designed to peer far out into space and far back in time and is capable of producing imagery of unprecedented clarity. Its observations over future decades are expected to bring new and challenging insights about the nature of galaxies, star systems, and intriguing objects such as quasars, pulsars and exploding galaxies. Most of all, the Space Telescope was designed to engage in a search for primordial galaxies as far as 14 billion light years away—near the time the universe came into existence—and to perform very deep red-shift studies to determine the large-scale structure of the universe. The Space Telescope was also designed with ten times the capability of the best Earth-based telescopes for distinguishing fine details in nearby star fields and planetary atmospheres. Orbiting 380 miles above the Earth, it is the greatest, best hope for obtaining definitive answers to questions about the structure of the universe on a large scale, including its size and movement. Unfortunately, after its launch in 1990, a defect was discovered in one of the telescope's mirrors, threatening disappointment to astronomers who had been looking forward to its data for nearly 20 years. While the telescope was still able to gather scientifically worthwhile views, its blurry vision fell far short of plans, able to see only about 3 billion to 4 billion light years away. Additional problems appeared when two of the spacecraft's six gyroscopes failed in 1991. The mirror was successfully repaired during a shuttle mission in December 1993, however, and with the new optics installed by astronauts, the Hubble can take photos of stars 10 to 11 billion light years away—close to its originally planned capability.

Scientists are optimistic that the repaired telescope will be able to start

his whole life became centered around measuring the far reaches of space and time.

As British spectroscopist Leonard Searle would later say, "Allan Sandage concentrates so incredibly fiercely. He's a marvelous scientist. He flings himself into things. He seems to be a man of passion."

On the basis of his examination of the spectral characteristics of certain globular clusters, Sandage would ultimately conclude, many years later, that they and the universe in general are less than 25 billion years old. Using the scale Sandage developed, astronomers were now able to measure distances to galaxies anywhere from millions to billions of light years away.

For an observational astronomer it was an exciting time. New evidence kept coming in constantly that astrophysicists and their colleagues in other disciplines used to explore questions such as: Do neutrinos exist? What happens when a star explodes? How do stars evolve? What happens deep inside them?

answering some fundamental questions about the universe, such as: What is the size and age of the universe, and how fast is it expanding? Is the universe open or closed—will it expand forever or will it, at some point, stop expanding and then collapse in on itself? How does matter evolve? And do black holes actually exist? The repaired Hubble will be able to measure the velocities of stars suspected of being sucked toward the centers of black holes, and this observation would go a long ways toward establishing the existence of black holes.

Another satellite, the Gamma Ray Observatory, carries four large "telescopes," some as large as a compact car, each of which recognizes gamma rays within a specific energy range. Because gamma rays, like all radiation, can only be detected when they interact with matter, the GRO detectors convert the rays to flashes of visible light, which are counted and measured. Gamma rays are the highest energy radiations in the electromagnetic spectrum, with a range from tens of thousands to tens of billions of electron volts (eV). (Visible light produces only a few eVs, by contrast.) Cosmic gamma rays cannot be detected at all on Earth because they can't penetrate the atmosphere. But many of the most interesting objects discovered in the past few decades, including quasars, pulsars and neutron stars, involve large releases of energy, which produce gamma rays, and astronomers hope to gain new insights on these structures and their dynamics through the data rounded up by GRO. Scientists even think that matter being drawn into a black hole gives off gamma ray radiation, in which case they might catch a glimpse of it just before it disappears.

WHAT HAPPENS INSIDE STARS?

Hans Bethe was a top-flight nuclear physicist who had studied with Arnold Sommerfeld in his native Germany, as well as Ernest Rutherford at Cambridge and Enrico Fermi in Rome. When Adolf Hitler came to power, Bethe left Germany for the United States, where he helped work on the bomb. But one of his greatest contributions to science lies in his insights about what goes on inside stars, which he worked out in 1938. He made use of his extensive knowledge of subatomic physics and of Arthur Stanley Eddington's conclusions that the larger a star is, the greater the internal pressure and the higher the temperature.

Bethe began with a hydrogen nucleus (a proton) and a carbon nucleus, which started a series of reactions finally resulting in the reformation of the carbon nucleus and the formation of a helium nucleus (an alpha particle). That is, the engine of a star ran on hydrogen fuel, leaving helium behind as "ash" and carbon as a catalyst. Because stars like our Sun are composed mostly of hydrogen, many have enough fuel to last for billions of years. Bethe also outlined a second possible scenario, in which hydrogen nuclei form a union directly (without the carbon catalyst) to form helium in several steps, a mechanism that could take place at lower temperatures. Bethe received the 1967 Nobel prize for physics.

In 1948, George Gamow became interested in Bethe's idea that nuclear reactions power stars and serve as the source of their radiant energy. Like Bethe, Gamow was also a physicist by training, although he had been interested in astronomy since his 13th birthday, when his father had given him a telescope. Born in Russia, Gamow studied in Europe at several universities, including work with Niels Bohr and Ernest Rutherford. Traveling to the United States in the 1930s, he collaborated with Edward Teller on theoretical work in nuclear physics, accepted a teaching position at George Washington University and decided to stay. Gamow worked out the consequences of Bethe's idea and found that as a star uses up its basic fuel, hydrogen, in this process, the star grows hotter. He hypothesized that instead of slowly cooling, our Sun would slowly heat up and eventually destroy life on Earth by baking it and maybe even engulfing it.

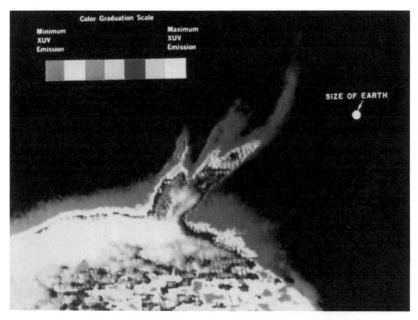

Image of an eruption on the Sun taken by the Skylab 3 *crew in August 1973. Notice how far this eruption extends into space only 90 minutes after it began.* (NASA)

Space-based studies of the Sun have confirmed Bethe and Gamow's ideas about fusion-powered stars, in addition to discovering many other facts about the Sun, including the existence of a solar wind of charged atomic particles that constantly rushes past the planets to the edge of the Solar System. On three missions of the U.S. Skylab space station in 1973–74, astronauts focused primarily on the Sun, sending back 75,000 photographs of solar activity, including six solar flares (explosive releases of energy from the Sun).

NEW METHODS, NEW DISCOVERIES

In 1931, radio engineer Karl Jansky, from Oklahoma, used an improvised aerial to locate sources of interference in radio telephone links and ended up founding an entirely new branch of astronomy, known as radio astronomy. He published his first results in 1932 and by 1933 had found that the celestial radio emissions he had discovered came from the Milky Way.

The field did not take off right away, though. The first real radio astronomy began in Australia in 1946 with solar studies led by E. G. Bowen. Then in 1947, radio astronomers homed in on the first radio object ever detected that they could match up to an object observable with the eyes: the Crab Nebula. Today, radio maps of the sky can be used to create images that help us "see" ranges of temperatures and heat divisions in far-off galaxies and stars.

A radio telescope gathers radio waves using an antenna or "dish," usually a scooped reflecting surface shaped like a satellite dish and made of metal or wire mesh. It's possible, though, to build a radio wave antenna without building a dish; Karl Jansky did so, as did Anthony Hewish and his colleagues at the University of Cambridge. The world's largest fixed-dish radio antenna makes use of a natural valley, located at Arecibo, Puerto Rico. Built in 1963, the telescope dish measures 1,000 feet in diameter, spanning some 25 acres. The antenna dish consists of 40,000 individual reflecting panels attached to a network of steel cables. The panels focus incoming radio waves from outer space onto a detecting platform suspended above the dish. In recent years, radio astronomers have increased the resolution of their equipment by constructing banks of antennas, such as the Very Large Array (VLA) recently completed in Socorro, New Mexico, the largest array of radio telescopes in the world. The VLA is made up of a total of 27 telescope dishes, arranged in a big Y shape in the middle of a flat, barren plain.

Radio astronomy didn't really begin to catch on until after World War II, but when it did, astronomers became excited about this new way of

STELLAR EVOLUTION:
CECILIA PAYNE GAPOSCHKIN

Among the great questions explored by astronomers in the 20th century is "What is the life history of a star?" Once it became clear that stars are born, burn brightly while young, grow old and die, astronomers rushed to unfold the details. Among the frontrunners in this process early in her career was Cecilia Payne Gaposchkin (1900–79), a woman who, by the second half of the century, became the best-known woman astronomer of our time and can readily be classed as one of the finest astronomers ever.

Cecilia Payne Gaposchkin
(Courtesy Yerkes Observatory)

searching the skies. Radio waves could penetrate the dust clouds in space that absorb sunlight and make optical astronomy difficult. Radio waves are especially helpful, therefore, in studying the center of our galaxy, which we cannot see at all by ordinary means.

Radio astronomy was the means used when Allan Sandage and Maarten Schmidt discovered quasars and when Anthony Hewish and Jocelyn Bell discovered pulsars.

Born Cecilia Payne in Wendover, England, she won a scholarship to the University of Cambridge in 1919, where she was inspired by A. S. Eddington to pursue astronomy. Harlow Shapley invited her to join his staff at the Harvard College Observatory, where she worked on spectra under Annie Jump Cannon, who oversaw the compilation and classification of Harvard's vast catalogue of photographic plates of spectra of stars. (Cannon alone classified a third of a million spectra using her own system.) At Harvard, Payne completed her Ph.D. thesis, "Stellar Atmospheres," which Otto Struve described as "the most brilliant Ph.D. thesis ever written on astronomy." At 24, Payne synthesized data from the spectra and her own observations to deduce the temperatures represented by each spectral type and the elements of which stars' atmospheres in general are composed. A gifted scientist, Payne Gaposchkin liked to describe herself as a field naturalist who was good at "bringing together facts that were previously unrelated and seeing a pattern in them." Although she tended to be unassuming and would not have thought of it that way, the trait she described is one of the key qualities of a great theorist.

In 1934 Payne married Sergei I. Gaposchkin, an expert in variable stars who had recently joined the Harvard College Observatory. Together they coauthored many papers. In the 1950s Payne Gaposchkin also authored three major texts on stellar evolution, *Stars in the Making* (1952), *Introduction to Astronomy* (1953), and *Galactic Novae* (1957). She became the first woman in Harvard's history to become full professor, a rank she achieved in 1956; she was also the first to become chair of her department, a post she held for 12 years. Her final work, published in 1979, was *Stars and Clusters*.

Like most people intensely involved with their work, scientists are sometimes plagued with attacks of jealousy when a colleague makes a breakthrough. To avoid feelings of jealousy, Payne Gaposchkin liked to say, she would suggest that scientists should repeatedly ask themselves whether they were thinking of the advancement of knowledge or their own career advancement. Payne, clearly, was more inclined to think about the advancement of knowledge.

Quasars

Several compact radio sources were discovered during the 1950s, but radio telescopes at that time could not pinpoint celestial objects very accurately, so it was difficult to match these objects up with visual images, using an optical telescope. One of these compact sources, known as 3C273, was blocked from vision by the moon in 1962 and its exact position could be established. Photos taken by Allan Sandage with the 200-inch Hale telescope atop Mount Palomar showed a dim, star-like object in that position. But this star had an unusual spectrum. It contained absorption lines that could not be identified. It, and later others like it, became called quasi-stellar radio sources (*quasi-stellar* meaning "star-like"), or quasars for short.

In 1963 Maarten Schmidt discovered that the absorption lines in the spectrum of 3C273 were common ones, but that they shifted toward the red end of the spectrum by an extraordinary amount. During the following years, astronomers discovered a large number of quasars with extremely large red shifts.

SCIENCE AT ITS WORST—AND BEST: COLD FUSION FEVER

The whole point of science is to find answers to the unknown, and all scientists work at uncovering answers both large and small. Many dream of making the spectacular breakthrough that will astound the world and make them famous. But in the latter half of the 20th century, from the viewpoint of those in other disciplines, physicists may seem to have had more than their share of cutting-edge discoveries.

So when two well-established chemists, B. Stanley Pons and Martin Fleischmann, saw the results of their experiments in March 1989, they were justifiably elated. They had, it seemed, done the impossible—and outdone physicists at their own game in the bargain. The next thing anyone knew, a press conference had been called, and everything spun out of control from there. They had found, they announced to the press from the University of Utah at Salt Lake City, a method for generating clean heat from nuclear reactions that take place at room temperature, and use only sea water as fuel.

Cold fusion, as this process came to be called, would mean cheap power—and could mean a wealth of possibilities for those who held its secrets. It was the alchemist's dream come true. The University of Utah, eager for the funds and stature that this discovery and further research could bring to the institution, promoted the idea of founding a national Cold Fusion Institute. The governor and legislature of Utah pledged $5 million for developing the idea further. Testimony before the House Science, Space, and Technology Committee resulted in the suggestion that $25 million in federal funds be set aside for Utah's cold fusion research.

The spectrum of a star can not only disclose its chemical composition, but from the Doppler shift, or red shift, one can deduce its velocity relative to Earth. Many systems of stars outside our galaxy have a shift in their electromagnetic spectra toward the red end of the spectrum, and astronomers accept this to be a Doppler shift, indicating the velocity with which these systems are moving away from us, a confirmation of the expansion of the universe. Red shifts caused by the expansion of the universe are called cosmological red shifts. If the red shifts of quasars are cosmological, the quasars must be at extraordinary distances—as much as a billion light-years away—making them the farthest objects ever observed in the telescope. Also, because they can be observed over such distances, their energy output

---

But all was not, after all, as good as it looked. Fleischmann and Pons, it turned out, had not run the controls they should have run. (A control in an experiment sets a standard of comparison for verifying the results.) Other laboratories were having trouble duplicating their results. (Another of the necessary facets of a scientific experiment: results must be repeatable.) Fleischmann and Pons said the other scientists weren't doing the experiments correctly, but they were unwilling to provide details. Of course, with so much money riding on their process, their reluctance to be open might be understandable. But in the end, they could not defend their results. Pressured, perhaps, by over-eager administrators at the university or by their own ambition, they had spoken too soon, before they could really be sure of their data. Somewhere, somehow, they had made a mistake. Fleischmann, a visiting professor of electrochemistry from the University of Southampton in England, was a distinguished scientist of considerable repute. Pons was a tenured professor at Utah. The reputations of both were badly tarnished, and Pons left Utah. The president of the University of Utah, who had told the press that cold fusion "ranked right up there with fire, with cultivation of plants, and with electricity," resigned. And the newly founded Cold Fusion Institute closed its doors. The fever was over.

The cold fusion flurry is, most of all, an example of how gullible we all can be. The idea of cold fusion fed the dreams of so many that it won the day, however briefly, against skepticism, underscoring the fact that, as Richard Feynman once said, "The first principle is that you must not fool yourself, and you're the easiest person to fool." The incident also shows that science as a whole is not easily fooled for long; science is a self-correcting process and sooner or later, when it has gone off on a wrong track, it discovers its own error, retraces its path and looks for better answers.

MILTON HUMASON:
BORN SKYWATCHER

Milton Humason (1891–1972), also an astronomer at the Mount Wilson Observatory, truly started out at the bottom of the mountain and worked his way up. He began as a mule packer who hauled equipment up the steep trail to the observatory, north of Los Angeles. Later he graduated to busboy in the observatory dining hall and then janitor in the observatory itself. But his love for astronomy won out. The first step was a chance to serve as assistant on one of the small telescopes on the mountain. He augmented his eighth-grade education with avid questions, reading and studying, and in 1928 he became Edwin Hubble's close collaborator.

Working under Hubble's guidance, then later on his own and with other astronomers at the Mount Wilson and Palomar observatories, Humason measured red shifts in the light of thousands of remote galaxies. Dedicated and skilled as an observer, he was expert at obtaining measurable spectrograms of faint, distant galaxies. His observations established what is still found to be the case today, that the Hubble Law holds true throughout the observable universe.

Milton Humason (Mount Wilson and Las Campanas observatories)

must be enormous. As Hubble had shown, velocity with which an object is moving away from us is proportional to the distance. This fed into the idea that the universe was created by a huge explosion and that the galaxies are the resultant debris flying in all directions. It also meant that the quasars were very, very far away.

The discovery of quasars caused great bewilderment among astronomers. The consequence of this discovery was that one either had to question the validity of the yardstick of the astronomer, the red shift, or agree that there are processes out there for which we have no explanation at all. Some quasars have been spotted that are thought to be more than 12 billion light-years away. Probably they are galaxies with extremely active centers but are so far away that they seem to be very dim single stars.

Not all astronomers, however, believe that quasars exhibit cosmological red shifts. For example, one American astronomer, Halton Arp, has discovered a number of systems consisting of a quasar and a galaxy that seem to be physically connected but exhibit very different red shifts in their spectra. He therefore argues that some still unknown mechanism other than the expansion of the universe must cause these red shifts. Most astronomers believe that quasars have cosmological red shifts and that the systems discovered by Arp only appear to be connected, and that they in reality are at very different distances from Earth.

A Bug in the Data

In July 1967, Anthony Hewish and his students had strung a long antenna in a field near Cavendish Laboratory in England to create a more powerful radio telescope for observing "twinkling," or scintillations, of radio starlight. It was the job of graduate student Jocelyn Bell to check over the charts each day to look for interesting data. In August she noticed a strange twinkling in a spot in the sky where ordinarily there would be nothing like that. Hewish thought it was probably noise in the receiver. They laughingly called this a signal from "little green men" and continued gathering data. Not only did that signal persist, but Bell found three more, similar pulsating radio sources. They began to realize that their data reflected a real phenomenon: a type of object that had never been detected before. They began to search for explanations in terms of known laws of physics.

What Hewish, Bell and their colleagues had discovered became known as pulsars (because they pulsate), and the scientists realized that they had detected neutron stars—a type of star that has become so incredibly dense that it may have a mass as large as our Sun's crammed into space the size of a mountain. Although it had been thought that neutron stars might exist, no one had ever detected one before.

THOSE WHO LISTEN:
THE SEARCH FOR EXTRATERRESTRIAL
INTELLIGENCE

A small group of dedicated, visionary scientists—including respected American astronomer Carl Sagan—has actively participated in what is known as the search for extraterrestrial intelligence (SETI). While the likelihood of "ETs" ever arriving here is slim (such a trip would require many generations to complete from most corners of the universe), many scientists think we could someday receive a signal from some Earth-like planet in another solar system somewhere in the universe. But when it arrives, we'll only recognize it if we're listening.

Searching for signals from civilizations beyond the edge of our solar system, though, could be likened to trying to find a needle in a cosmic haystack, or straining to hear a cricket's voice against the roar of Niagara Falls—wondering all the time if there *is* a needle, or a voice. Or if, in the end, we'll discover nothing more than a pile of haphazard scrap, or the relentless persistence of unsilent nature.

But some questions seem eternal, rooted deep in human consciousness. "Are we alone?" is such a question, and in the 1980s and 1990s, the tools of technology have come of age enough to enable well-defined experiments aimed at finding the answer. Worldwide, scientists have begun sifting through the wide range of signals from space in a well-organized search.

As astrophysicist and SETI scientist Bernard M. Oliver remarked in a 1986 interview, "If we are right, then over the last few billion years a great number of intelligent civilizations have grown up like islands in this galaxy, and it's inconceivable to me that all of them would go through their entire histories in isolation."

The first great breakthrough in SETI methods came in 1959 when two scientists, Philip Morrison and Giuseppe Cocconi, suggested that radio astronomy could be used to communicate with other worlds.

Why radio astronomy? As Oliver explained it, "In the interests of economy and efficiency, a message carrier should meet the following criteria: (1) the energy . . . should be minimized, other things being equal; (2) the velocity should be as high as possible; (3) the particle should be easy to generate, launch and capture; and (4) the particle should not be appreciably absorbed or deflected by the interstellar medium."

These criteria are easily satisfied by radio waves. They are fast, efficient and relatively cheap. Therefore, it seems only logical that intelligent civilizations (if they exist) would choose the radio spectrum for sending long-range signals across the vast distances of interstellar space.

The first radio-telescope search for extraterrestrial intelligence was a project called Ozma, set up in 1960 by SETI pioneer Frank Drake at the National Radio Astronomy Observatory (NRAO) in Green Bank, West Virginia. Drake chose two nearby solar-type stars, Tau Ceti and Epsilon Eridani, and spent a total of 150 hours "listening." He found nothing, but he had made a beginning.

Since Ozma, more than 30 SETI experiments have been launched in the United States, the former Soviet Union, Australia and Europe—all without results. Still, despite many hours of combined listening time, they have covered only a tiny slice of the total range of possibilities. Nearly endless combinations of direction, segment of the spectrum and types of signal modulation remain to be explored.

By the 1980s and 1990s, new technology and the enormous data-handling capabilities of today's computers had revolutionized the search. Multichannel analyzers now can swallow in literally millions of radio channels at one time. One project, established by a nonprofit group called the Planetary Society, inaugurated a new 8.4-million channel analyzer in September 1985. NASA's SETI project, which lost funding from Congress

Astronomer Jill Tarter's daughter used to list her mother's occupation as "looks for little green men." But for Tarter, SETI is both serious and exciting science, and she has spent most of her professional career finding more and more technologically sophisticated ways to "listen" effectively for signals from other worlds. (Photo by Seth Shostak, SETI Institute, Courtesy, NASA Ames Research Center)

in 1993 but has since revived thanks to private funding, uses VLSI (very large-scale integrated) circuitry to pull in as many as 10 million individual frequency channels. Real-time data from the telescopes is analyzed by this equipment for significant data, which is passed on to a signal analyzer that in turn passes it on to a powerful computer.

Without this automated sifting process, the volume of information from the radio telescopes—billions of bits of digital data—would be impossibly daunting. Before the more advanced technology was available, one five-day observation period could produce 300 tapes of data, or more. Analyzing the data on those tapes, in turn, could take as long as two and a half years to pump through a computer. After one such experience, astronomer Jill Tarter drew a cartoon of a pair of feet sticking out from beneath a mountain of computer printout, which she labeled "Buried alive!" "You can't do it in non-real time," she would later explain. That is, if you don't process the data as it comes in—in real time, you'll find "you can't store it, and you can't process it with human intelligence. You've got to have something that is far more dedicated and, once you've told it the rules, plays by the rules consistently and with complete concentration." The task requires equipment that's able to weed out the vast quantities of irrelevant noise and retain only the signals that might be interesting. The new technology does just that.

But having equipment to analyze the data is only part of the challenge. Given the vastness of the universe, where do you look? And what do you search for? SETI researchers usually use one of two approaches: home in on a few likely stars using sensitive equipment, or use less sensitive

IN THE BEGINNING . . .

For centuries people have wondered how the universe began and how or whether it might end, and the last half of the 20th century is no exception. But most astronomers in the 1950s tried to sidestep the question when it came to trying to explain it scientifically. George Gamow was an exception. In 1948, Gamow worked out a scenario that would result in the formation of the various elements of the universe from an explosion of the kind of original "cosmic egg" or "superatom" suggested by Lemaître. A widely read popular writer about science, Gamow soon had a widespread reputation for this idea about how the universe began, although it was by no means universally accepted.

In fact, many scientists felt that the question of "the beginning" either was outside the realm of science or was offensive. British physicist Fred

equipment in a broader range of frequencies to conduct a broad, full-sky sweep. It is a vast task, one that may have no positive results within our lifetime. And it's a project with a lot of "ifs." For one thing, would extraterrestrial civilizations, if they indeed exist, be likely to follow the same reasoning, choose the same frequencies and beam in our direction? Perhaps even more important, would they make this costly attempt to reach us? Or would extraterrestrial scientists have trouble convincing some galactic congress to provide funding for such an ambitious, and possibly useless, venture? Also, given the billions of years the universe has existed, this moment when we have reached the ability to search the cosmos for signals from other civilizations may not coincide with the time when signals would arrive. A signal coming from four light years away would take only four years to arrive here. A signal sent today from a civilization 100 light years away, though, would not reach us for another 100 years.

But whether the answer is 10 years away, 20, 50 or even 100, most searchers agree that we will learn, not just from the answers that we may or may not find, but from the ways in which we seek.

And so the search goes on. Is intelligence an isolated phenomenon, and Earth its only voice? Are we a short-lived quirk of nature, or part of a greater cosmic community? Can civilizations such as ours endure long enough to reach outward toward other worlds, or is humankind doomed to perish, alone and unheeded, a dead end in the handiwork of nature? Are there other voices out there like our own, other searchers in the cosmic darkness, seeking light and companionship? Seeking hope? Only time— and SETI programs throughout the world—will tell.

Hoyle, also a popularizer of science, numbered among these, and he came out with a competing theory, with Austrians Hermann Bondi and Thomas Gold. Incensed at the idea of an explosive beginning of time, they painted a picture of what they called the "steady-state universe," in which matter is continually being created that in turn powers the expansion Hubble observed when he saw that all galaxies are moving away from us. Hoyle poked fun at what he laughingly called Gamow's "Big Bang" theory. Imagine his amazement when the name caught on.

The Microwave Background

Meanwhile, in 1964, two researchers at Bell Laboratories in New Jersey— Arno A. Penzias (1933–) and Robert W. Wilson (1936–)—were using the lab's big radio antenna to look for weak signals from the sky. But they were having a hard time getting rid of the background noise in order to get a clearer signal. They took the equipment apart. They checked the dish. They checked all the connections. They even found pigeons nesting in the dish, and so they carefully removed the birds and took them miles away to nest somewhere else. The pigeons came back. They removed them again. Nothing would get rid of the microwave background noise the scientists were hearing. Slowly it dawned on them that the signal they were hearing was coming from space.

Penzias and Wilson had discovered a fundamental fact of nature: a constant microwave background radiation that permeates the universe. Like a great echo, this radiation seemed to imply some great past event that had raised the temperature of the entire universe at a time so long ago that by now it has almost, but not quite, entirely dissipated. It was the first real reinforcement for the idea proposed by young Gamow in 1948, who not only predicted this radiation, but figured its exact temperature correctly as 3°K above absolute zero.

Then in 1992, another piece of evidence came to light. The science team working on data beamed back to Earth by the Cosmic Background Explorer (COBE) spacecraft announced that, unlike previous evidence, the new data indicated that the cosmic background radiation has "ripples." Always before, the data gathered by Wilson and Penzias and all other successive researchers had indicated that the temperature of the background radiation was the same, no matter where in the sky you looked. From this constant temperature, scientists had deduced that the early universe must have been smooth and uniform, completely lacking in what cosmologists now refer to as "lumpiness." Wherever we look in the skies we see clumps of matter—galaxies, nebulas—interspersed with empty space. This spottiness is what cosmologists mean when they talk about lumpiness.

The new data is compelling because of its sheer quantity. COBE, launched by NASA in 1989, made several hundred million temperature measurements during the first year in orbit alone. In this enormous pool of data, the COBE team discovered tiny variations in temperature, amounting to only 30 millionths of a degree warmer or cooler, in areas where minute fluctuations in the density of gas occurred in the early universe, about 300,000 years after the Big Bang. (When we talk about a time only 300,000 years after the beginning of time, it's as if, in human terms, we spoke of the first day in the life of a 90-year-old person.)

As the universe expanded, these early areas of temperature fluctuation also grew, so that now, the areas detected by COBE are as large as billions of light years across. They are so large, in fact, that they could not be precursors to even the largest clusters of galaxies we have observed. But their discovery made scientists confident that more density fluctuations of smaller sizes will turn up. And they seem to back up the "inflationary" model of the birth of the universe.

Of course, whether the universe began with a Big Bang or otherwise, the idea of a creation event always raises the question, where did the matter come from in the first place? But this, as cosmologist Stephen Hawking once said, is like asking what is located five miles north of the north pole. Or, put another way, any question about "before the Big Bang" is unphysical. Time exists inside the Universe—the Universe does not exist inside time.

Black Holes

Meanwhile, in the 1950s at the University of Cambridge, a very personal battle was taking place. A young graduate student named Stephen Hawking had just received a chilling piece of news—a diagnosis explaining why, in the past few years, his gait and speech had become less and less coordinated. He had what is commonly known as Lou Gehrig's disease (amyotrophic lateral sclerosis), a degenerative paralyzing affliction that would put him in a wheelchair within a few years. In the coming years he could look forward to increasing physical disabilities and, eventually, death. The brilliant young student went into immediate depression. Why pursue the promising career that he had begun? Why go on at all? Several months went by and he made no progress on his work.

Stephen Hawking
(Courtesy of the Archives,
California Institute of
Technology)

Although Hawking's health could not be recovered, luckily for science, his career could. His adviser came up with a plan: Toss a problem into the young man's lap that was so fascinating that he couldn't resist it. That was how Hawking got started looking into the finer points about black holes and became the most world-renowned expert on the subject, one of the most provocative in modern astronomy.

American physicist John Archibald Wheeler (1911–) coined the term *black hole* in the 1960s to describe the structure formed when a star collapses in upon itself and, ultimately, becomes no larger than a single point in space, a singularity. When this happens, according to Einstein's theory of relativity, nothing can escape the enormous concentration of mass—not even light. And so, a black hole cannot be seen. No one has ever located a black hole by any other means, either, but physicists and astronomers are sure they must exist in the universe, and there is good reason to believe that a location known as Cygnus X1 is a black hole, based on signals from the surrounding region that indicate the presence of an object even though nothing can be seen.

In 1974, Hawking came up with the concept that "black holes aren't black"—that is, he thinks that black holes can slowly bleed radiation. Possibly, he says, black holes may evaporate like snowballs in the sun. This seems to be a contradiction in terms, since a black hole, by definition, is so massive that nothing can escape its gravity, not even light. That's why it's called a black hole. The edge of the black hole, called the event horizon, allows nothing back out.

But Hawking has pioneered the application of quantum mechanics to black hole theory, and he has come up with the idea that matter can be created in "virtually" empty space at the event horizon. That is, according to quantum theory, virtual particles can wink in and out of existence so fast that they don't disturb the balance required by the laws of conservation of energy and matter. Hawking thinks that this can happen at the event horizon of a black hole and that, while most virtual particles would immediately wink back into the black hole, occasionally some would bleed off in another direction, slowly leaking radiation away from the black hole.

This idea fits well with the most recent contribution to theories about how the universe began, proposed in the 1980s by Alan Guth of the Massachusetts Institute of Technology. Called the inflationary model, Guth's theory proposes that during the first fraction of a second of the existence of the universe, the entire universe expanded enormously in a sudden, ultra-brief (one-trillionth of a second) burst, expanding the universe from the size of an atom to billions of light years across.

Missing Mass

One of the eeriest phenomena observed in astronomy—rivaling black holes—was introduced by Swiss astronomer Fritz Zwicky (1898–1974) in 1933. The phenomenon is sometimes referred to as "missing mass," because many clues tell us something is there that we cannot detect with telescopes of any kind. Zwicky called it *dunkle Materie,* or dark matter. Zwicky was an irascible, renegade character, the kind of person that "likes to prove the other guy wrong," as one colleague put it, and it was not surprising that he was one of the first to notice that the universe's ledger books didn't balance. He came up with the idea of "missing mass" to describe the dark, invisible mass that must be there—something between 10 to 100 times the amount of matter we can detect in the galaxies and between galaxies.

No one listened to him at first, but by now some astronomers think that the matter we see could not be sufficiently massive to cause the coalescence into the stars and galaxies we see. There seems to be more gravity in the universe that we can account for based on observations of luminous matter.

In fact, the current suspicion is that we have found no way yet to detect as much as 99% of the matter in the universe. It does not give off light and is not detectable with X rays or radio astronomy. But we see its gravitational effects.

For example, in the 1970s Vera Cooper Rubin (1928–) and W. Kent Ford at the Carnegie Institution of Washington, D.C. gathered extensive

Fritz Zwicky (California Institute of Technology)

Vera C. Rubin (Carnegie Institution of Washington, D.C.)

data indicating that in clusters of galaxies, the galaxies far from the center move at too high a speed for the gravitational effects of the mass we see—stars and glowing gas—to hold them within the clusters. Their investigations of the rotation of spiral galaxies revealed the presence of a dark halo around the edge of the galaxies. Within this halo, Rubin and Ford estimate, resides something like 10 times the amount of matter we can actually see in the galaxy. Statistical analysis of the motions of galaxies on a larger scale than clusters backs these ideas up with an estimate that dark matter constitutes 30 times the amount of matter we can detect (except within galaxies, where ordinary matter dominates). Since 1978, a group at Carnegie including Rubin, Ford, David Burstein and Bradley Whitmore has analyzed more than 200 galaxies with confirming results.

What is this dark matter? No one is yet sure; there may be more than one kind. But it is not at all likely that it is mostly ordinary matter—that is, it's not mostly black holes that have formed from stars or dim stars we can't detect or small chunks of solid matter. By 1984, the candidates were neutrinos (if mass greater than zero) or hypothetical particles: photinos, gravitinos and axions. Some scientists think dark matter may be slow-moving elementary particles left over from the big bang.

As British solar astrophysicist Martin Rees remarked, "The most clear-cut way to settle the nature of the hidden mass would of course be to detect the objects that make it up."

Some astrophysicists suggest a connection between brown dwarfs and missing mass, since a brown dwarf is a theoretical "star" formed when a lump of gas contracts but has a mass so small that nuclear reactions do not begin in its core. Because there are no nuclear reactions taking place, a brown dwarf would be very dark and difficult to detect. No one has ever found a brown dwarf. Are they really there? Are they part of the missing mass?

But still not everyone is convinced that dark matter even exists. In 1986 a team of seven astronomers noticed that our galaxy, the galaxies of the local group, and other components of the local supercluster of galaxies move in a particular direction, toward a point in the direction of the Southern Cross, which they called the "Great Attractor." This yet unknown force, they pointed out, could account for the movements detected by Zwicky, Rubin and others, in which case the missing mass explanation would then be unnecessary. But if so, then what is this strange force, the Great Attractor? And what dynamics are involved?

PULLING IT ALL TOGETHER . . . ?

Most intriguing of all, the recurring question that haunts cosmologists and subatomic physicists alike is, How did it all begin?

Nearly all theorists now agree that there must have been a beginning and that it must have been a big bang, a cataclysmic explosion originating from a particle smaller than the smallest nucleus of any atom. The characteristics described by Guth's inflationary theory seem incredible—but most theorists think Guth must be right because his theory explains so many of the characteristics of today's universe.

Why, for instance, didn't everything just blow apart so violently in those first few fractions of a fraction of a fraction of a millisecond that all matter was evenly dispersed far and wide, with no coalescing objects, no lumpiness, no stars, planets, galaxies and comets? Or, if not enough momentum was available to blow everything wide open, then why didn't the closely packed infant universe just crush in upon itself? Theorists have actually calculated what the critical value would have to have been for the universe to have avoided either fate, and it works.

Many theorists have tried to figure out exactly what happened in those first few moments, though, trying to reach back in time to a point when the atom's structure was simpler, looking for the simple symmetry behind the complexities we see today. They have looked back to a time when the newborn cosmos for the briefest instant, in one of its earliest, earliest stages, reached temperatures above a billion billion billion degrees, and in this brief moment, the weak, electromagnetic and strong forces were all still just one force. These efforts are the GUTs—grand unified theories—and so far, none of them has worked. Sheldon Glashow once remarked that GUTs are neither grand nor unifying nor theories. So, as particle physicist Leon Lederman puts it, a better term for them than *theories* would be *speculative efforts.*

To carry this force an ancient messenger particle has been postulated, the X bosons (and anti-X bosons), super-massive particles that played the part of messenger for the grand unified force that photons play today for the electromagnetic force. At this time, particles and antiparticles could transform themselves rapidly from one to the other. For example, a quark could become an antielectron (positron) or an antiquark. An antiquark could become a neutrino or a quark. And so on. This process was mediated by the X bosons.

Even more ambitious, also highly speculative, are the efforts to bring together all four forces, including the gravitational force, combining quantum chromodynamics—the behavior of quarks and their color properties—with gravity. Sometimes referred to as TOEs (theories of everything), these concepts become very complex. One, known as Superstring theory, gained great popularity for a while, explaining that in the very earliest flicker of big bang cosmology, there were no point particles, only short segments of string. Ten dimensions are required to make it work—nine space and one time. To explain why we only are aware of three space dimensions, theorists explain that the other six dimensions are curled in on themselves.

Many scientists are nervous, though, about so much theorizing, building complex theories upon complex theories with little or no chance of subjecting the theories to experiment. And the study of smaller and smaller elementary particle structures requires larger and larger accelerators and detectors. The final goal—to understand the fundamental structure of matter through unified symmetry principles—is commendable, but the mathematical complexities required are immense.

As a result, many critics worry that cosmological science is in danger of becoming a new kind of mythology, with little possible connection with testable science.

Overall, however, the past five decades have offered astronomers and cosmologists a vast richness of new data, new vistas, and new ideas. Researchers have developed far more sophisticated ways of investigating the universe, its structure and origins, making possible a far greater degree of precision and accuracy than ever before. And a highly fruitful cross-fertilization with particle physics has produced important new insights in both fields.

As a result the universe has become a vastly larger and more complex place than ever before—and ever more intriguing.

CHAPTER 4

SIGHTSEEING IN THE
SOLAR SYSTEM

While cosmologists envisioned the violent first moments of the universe and explored the consequences of evaporating black holes, the last half of the 20th century also offered up a treasure trove of new information for those interested in our neighbors in the Solar System, their history, structure and origins.

In 1957, when *Sputnik I* first began its lonely orbit around the Earth, we became tourists in the Solar System for the first time. Satellites orbiting the Earth and rocket probes into our atmosphere began to give us new views of our planet and its atmosphere—from outside ourselves. On their heels followed missions to the Moon (both with and without people aboard) and unmanned missions to nearly every planet in the Solar System; only Pluto did not receive a visit. What had been tiny, far-away specks focused faintly through the lenses of even the largest telescopes, now became rocky surfaces pockmarked with craters, swirling globes of glowing gases, volcanically active nuggets of turbulence, and icy deserts. In the vast silence of space, eerie robots whirred near them to send us close-up photographs and data readings that stunned everyone. Nothing was what it had seemed.

Spectacular, brightly colored close-ups of Saturn's rings and the volcanic eruptions on Jupiter's moon, Io, have become so commonplace today that we forget how far away these huge, spinning masses of rock and gases really are—and how impossible such views were to obtain just a few years ago. No one even suspected that Saturn's rings existed until Galileo spied evidence of them in his telescope in the 17th century. Now we have extensive data about their structure, size, movement and relationship to the planet and its moons, thanks for the most part to data gleaned by a spacecraft named

Pioneer 11, which flew by in 1979, followed in the 1980s by *Voyager 1* and *Voyager 2*.

Planetary science, which had always, since the time of Galileo, depended upon instruments—ever bigger, more complex and more expensive—now also had an even bigger price tag. The rewards were also vastly larger. We could actually see the surface of these far-off worlds, measure their atmospheres and study their histories. What we have learned about the way the universe works—from the dynamics of atmosphere and meteorology to the presence in space of a solar wind and radiation belts around our planet—not only adds to our basic store of knowledge, but provides priceless lessons about our planet, Earth.

THE MOON: CLOSEST NEIGHBOR

Locked together in a close cosmic dance, Earth and Moon, planet and satellite revolve around a common center of gravity, the larger one, green-blue-white, bursting with color, water and life; the other, smaller by nearly three-quarters, colorless, scarred, cratered and marked by time. Its hidden

Geologist Harrison Schmitt scoops up samples from the Moon's surface during the Apollo 17 *mission in 1972.* (NASA)

face, dark and pitted, is turned perpetually away from the Earth, toward the stars. What catastrophe occurred in the long-distant past to leave the Moon so cold and desolate?

At once our most familiar and most mysterious neighbor in space, the Moon has always intrigued the human mind. It is the subject of countless ancient mythologies, legends and songs. It was the calendar-keeper of the ancients, who depended on its regularity of motion across the sky and changing phases to count the passage of the seasons.

For scientists, the questions were many: What was the surface of the Moon really like? Where did it come from and how did it come into being? What was its geology? What was the face like that always turns away from Earth and had never been seen? When the United States and the Soviet Union developed the ability to leap beyond the confines of Earth's gravity, the Moon was the most natural place to first go exploring. Like the telescope and the camera, the rocket gave us a new tool that could help unravel the Moon's many secrets.

Between 1958 and 1976, the United States and the Soviet Union sent 80 missions to the Moon, although only about 49 of them completed their jobs as planned, and some never really got close. But those that made it—orbiters, soft landings, photo sessions, probes, two astronaut flybys, and six separate astronaut landings—brought or sent back a wealth of information about our closest neighbor in the sky.

We learned that the Moon is approximately the same age as the Earth: about 4.5 billion years. The relative abundance and proportions of oxygen isotopes suggest that the two bodies were formed nearby each other, although theories vary about the exact formation process. Did the Moon "spin off" the Earth, like a blob of paint? For this to work, Earth would have had to spin very fast, but the idea has had some supporters. Did it coalesce or condense out of the same planetesimal (pre-planet formational material) as the Earth and at roughly the same time? If so, why two "planets" instead of one? Or did Earth "capture" the Moon into its gravitational field as the Moon was passing by? Or maybe, during the early days of the Earth's formation, while it was still molten and was slowly building up out of many smaller planetesimals, a large "leftover" planetesimal smashed into it, vaporizing part of both the Earth and the planetesimal and spewing the molten debris into orbit around the Earth, where it eventually became the Moon.

Perhaps the most important legacy of our lunar exploration was a greater understanding of the Moon, a greater appreciation for our Earth and a key to some of the mysteries of our Solar System. In looking closely at the Moon, we have seen a world barren and desolate, too small and with too light a mass to hold on to an atmosphere. Unlike the Earth, the Moon is a world stunted and locked into its own past without growth or change—a world now so static (until the next impact) that footprints left on its dusty surface

GERARD PETER KUIPER:
PLANETARY SCIENTIST (1905–73)

Planetary science received a tremendous boost from NASA's programs, and astronomer-planetary scientist Gerard Kuiper was among the early visionaries who helped build a program from the beginning that could provide useful information to scientists.

Born and educated in the Netherlands, Kuiper left his homeland for the United States in 1933, when he joined the staff at Lick Observatory, near San Jose, California. There he studied binary stars and made spectroscopic searches for white dwarf stars. Later, at Yerkes Observatory in Wisconsin and its southern outpost, McDonald Observatory in Texas, Kuiper made contributions to the theory of the origin of the Solar System.

After World War II, Kuiper pioneered in the new field of radio astronomy, making spectroscopic observations of the planets and late-type stars. Using these techniques, he succeeded in discovering methane in the atmosphere of Saturn's giant moon, Titan, and carbon dioxide in the atmosphere of Mars, as well as water vapor in the Mira variable stars (located in the constellation Cetus). All of these discoveries served to tweak the curiosity of those interested in knowing if life existed or had ever existed beyond the atmosphere of our own planet. In addition, Kuiper discovered a new satellite orbiting Uranus and another spinning around Neptune.

In addition to serving as director of both the Yerkes and McDonald observatories, Kuiper later founded and directed the Lunar and Planetary Laboratory at the University of Arizona. For NASA, he served as principal

---

may last a million years. It's a kind of fossil that tells the history of our Solar System and our own planet. Our journeys to the Moon have already taught us much about ourselves and the other members of our biosphere—and the rarity of our existence.

For the next step, naturally, we turned to explore the nearest neighboring planet, Venus, the second planet from the Sun.

VEILED VENUS

Often seen brightly gleaming in the early morning sky, or glowing in the early evening dusk, Venus is our closest planetary neighbor. Known as Earth's "twin," it is close to the same diameter, size and density, and, only

Gerard P. Kuiper in about 1949, outside the Yerkes Observatory (Courtesy, Yerkes Observatory)

investigator (in charge of scientific responsibilities for a mission) for the Ranger missions to the Moon. He also made important scientific contributions to early NASA planetary exploration.

26 million miles away, it travels a similar orbit around the Sun. But, curious as we naturally were, even the most powerful optical telescopes of our biggest observatories could not penetrate the thick veil of clouds that surrounded the planet and blocked our view.

Speculation ran high among science-fiction writers, astronomers and planetologists alike. Maybe underneath the clouds Venus was a rainy, tropical planet, with seas and jungles teeming with life. We couldn't see the surface—but surely clouds meant water vapor and water vapor probably meant life, possibly even intelligent life. The idea died hard. From the 1930s on, radio astronomy and spectroscopy began to give hints that the atmosphere and temperature could not harbor life as we know it, and by 1961, microwave astronomy gave us further insights about the planet's direction and rate of rotation, atmospheric temperature, density and pressure, and a rough idea of topography. But only with the first planetary missions did we begin to confirm fully how unlike Earth Venus really is.

Between 1961 and 1989, more than two dozen U.S. and Soviet missions set off for Venus, most of them successful. Probes tested upper, middle and lower atmosphere, analyzing chemistry, cloud movements, pressure and temperature. The Soviet Union sent several landers that sent back pictures, from beneath the cloud cover, of the rocky plains of Venus. They relayed back surface temperature readings and more atmospheric analyses. And orbiters mapped the surface.

The U.S. *Mariner 10* spacecraft, launched in 1973, flew past Venus in early 1974 on its way to Mercury, and it made some interesting observations at Venus, including photos that were 7,000 times better than any we could take from Earth.

In 1978, the United States sent two Pioneer spacecraft, *Pioneer 12* and *Pioneer 13*. The second contained four atmospheric probes that deployed in different parts of the atmosphere and took temperature and pressure readings, plus data on wind patterns. But *Pioneer 12* was the stalwart hero of Venusian research. Designed to orbit Venus for 243 days, *Pioneer 12* continued to transmit scientific data for another *10 years*. By 1988, *Pioneer 12* had sent home more than 10 trillion bits of data, including a description of the extreme greenhouse effect which has trapped the Sun's heat to drive the planet's surface temperature as high as 900°F.

Between 1978 and 1983, the Soviet Union sent six Venera probes (*Venera 11* to *Venera 16*) in pairs, most of them landers. All sent back valuable details about the surface and atmosphere, including the first color pictures of the Venusian surface, analyses of drilled soil samples, and seismic experimentation. Two more missions in 1984, called *Vega 1* and *Vega 2*, carried experiments from several nations.

From all this data, a new picture of Venus had begun to form, when the United States sent the Magellan spacecraft to Venus in 1989, arriving in August 1990. Despite some technical difficulties, Magellan sent back stunning radar images of the surface and completed a 243-day radar mapping cycle during one complete rotation of the planet. Magellan revealed a surface torn by tectonic forces, ravaged by searing winds, its crust violently deformed and broken by giant volcanic eruptions.

Overall, the picture of Venus is a far cry from the imagined green, bountiful planet. At the surface a dim, peachy light filters through the thick Venusian clouds of deadly sulfuric acid. Gentle breezes stir the intense 891°F heat of the barren, dusty high-plains desert. At 217 miles an hour, winds in the upper atmosphere whip clouds overhead 60 times faster than the strange, backward rotation of the planet. In the distance, turbulent flashes of lightning brighten the sky around the conic shape of a volcano spewing out smoke and ash. As one technical memorandum from the U.S. *Mariner 5*

THE GREENHOUSE EFFECT ON VENUS

A victim of the so-called greenhouse effect, Venus's fiery atmosphere is the result of this close proximity to the Sun and its inability to partially disperse its tremendous "heat load" back out into space.

All planets absorb energy from the Sun in the form of solar radiation. This energy (which we call sunlight) is then reradiated outward as longer-wave infrared radiation. The surface temperature of a planet is determined by the balance between the sunlight it absorbs and the infrared energy it emits. The temperature rises if a planet absorbs much more radiation than it gives off.

On Venus the cooling process is particularly hindered by the thick Venusian atmosphere, which acts something like a one-way gate, permitting the visible solar radiation to enter, but preventing the longer-wave infrared radiation from escaping. The atmosphere functions like the lid on a pot, trapping the emitted energy and causing the planet to heat up.

Many scientists fear that an increase of carbon dioxide in the Earth's atmosphere, caused by continual burning of fuels such as gasoline and diesel, may also result in a greenhouse effect like our neighbor's and that a gradual accumulation of trapped "heat" may dramatically change the Earth's climate and ecological balance.

---

science team read: "Venus appears to offer roasting heat, a choking atmosphere, crushing pressure, and murky skies, to which forbidding weather and hostile terrain may perhaps be added."

SCORCHED MERCURY

Like a tiny moth circling a bright light, the planet Mercury wings closest of all to the Sun. Darting swiftly along its elliptical path, it defies the optical mirrors of our telescopes and quickly escapes our curious gaze with its forays close to its too-bright neighbor the Sun. Its surface parched by billions of years on this course, Mercury hangs poised, it seems, on the brink of fiery disaster.

But in reality, Mercury is a heavy mass of rock and iron baking in the Sun's blazing heat. Once every 88 Earth days it completes its path around the Sun, at an average of 58 million miles from the solar furnace. As a consequence, it is one of the most difficult planets of all to observe from Earth, viewable only for a short time in the morning and evening. Even the invention of the telescope did little to bring Mercury into better focus.

Mariner 10 arrived at Mercury in March 1974. Fitted out with special protection against solar radiation, the spacecraft flew within 437 miles of Mercury's surface to take a close look. Carrying two television cameras with 5-foot telescopes, an x-band radio transmitter, infrared radiometer, and equipment for ultraviolet experimentation, it transmitted nearly 2,500 pictures back to Earth. After zipping across the Mercury sky, the spacecraft caromed around the Sun to return twice more for photo sessions in September 1974 and March 1975.

As the first pictures came in from Mercury, scientists were struck by the planet's close resemblance to the Moon. From the evidence strewn across its parched and pitted surface, they concluded that Mercury's geological history may have been similar in many ways—the scene of thousands of meteorites smashing against it, some 3.9 billion years back in time.

In fact, the Moon is smooth compared to Mercury. With the exception of one gigantic flat area known as the Caloris Basin and a few other small patches, Mercury is almost entirely covered with craters. And, while scientists believe that the Moon's maria (vast flat, dark areas on its surface) were created by lava flows, most hold an entirely different theory about the formation of the Caloris Basin. Some evidence points to past volcanic activity on the planet, including some partial infilling of the basin itself, but the great Caloris Basin was probably formed by the most dramatic and important event in Mercury's history, collision with a huge asteriod.

From the evidence, planetary scientists surmise that an asteroid—possibly measuring more than 60 miles across—collided long ago with the tiny world and leveled an area nearly 850 miles across. As it smashed into Mercury at the speed of more than 315,000 miles per hour, this huge missile forever changed the face of the planet, pushing up mountain ranges over a mile and a half high around the rim of the crater it formed. Even a confused, jumbled surface on the opposite side of the globe, some specialists believe, may also have been created by the collision of the impact's gigantic shock waves as they circled the planet.

And that, many planetary scientists think, may have been the last important event in the planet's evolution. Judging from crust fractures and other evidence seen in *Mariner 10*'s photos, they deduce that Mercury had shrunk substantially prior to the impact, possibly due to the cooling of the iron core or a slowing of the planet's spin. After that, the planet evidently ceased to evolve. Mercury apparently "died" at the end of the same period of great bombardment that devastated the Moon, some 3.9 billion years ago.

Like a snapshot, the primordial state of Mercury's surface takes us back in time to glimpse details about the evolution and origins of our Solar System. This is the stuff of which progress in planetary science is made.

MARS, THE RED PLANET

Of all the planets in our Solar System, Mars has long been a favorite among science-fiction writers and astronomers alike. From the first days of telescopes it was the only planet whose surface features could be seen, and in the 1890s it inspired widely respected speculation—stimulated by the observations of astronomer Percival Lowell—that huge artificially-made canals criss-crossed its plains, signs of a present or ancient civilization. Seasonal changes in a periodically shrinking and growing ice cap at the north pole encouraged a vision of a planet much like ours, perhaps even with a growing season and a barren winter.

So the first time investigators succeeded in sending a robot spacecraft flying by—the U.S. *Mariner 4* in November 1964—the results were disappointing. From a distance just 6,000 miles above the surface, we saw our first glimpse of a planet that appeared to be flat, featureless, and lacking in life forms. Two more Mariner flyby missions in 1968 did little—except to hint at volcanic action and erosion—to brighten the prospects about what we might find on the surface of Mars.

Then a lonely robot observer called *Mariner 9* entered an orbit around Mars on November 13, 1971, becoming the first human-built object to orbit another planet. Despite a dust storm shortly after the spacecraft's arrival, the Mars of *Mariner 9* was a geologist's paradise. Far from being an ancient inert world, Mars proved to be a world where once rivers may have flowed and volcanoes erupted—a planet whose surface temperature and atmosphere may once have been capable of supporting at least some very simple form of life. Theories flew fast and furiously.

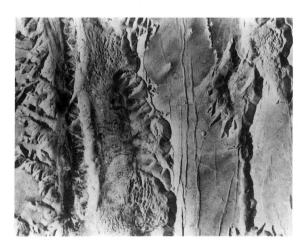

The Valles Marineris, as seen from the Viking 1 *orbiter. Markings in this vast valley have led geologists to think that the surface of Mars may at one time have experienced floods of water, by now long since evaporated.* (NASA Jet Propulsion Laboratory)

A volcano, which scientists named Olympus Mons (Mount Olympus), rose to more than 79,000 feet above the Martian surface—the largest known mountain in the Solar System. Its base was over 350 miles across, large enough to completely cover the state of Missouri! A huge canyon, of equally giant proportions—Valles Marineris—extended nearly a sixth of the way around the planet, a distance of nearly 2,800 miles long. It was 370 miles wide and nearly 13 times longer than the Grand Canyon (as far as the distance from New York to California). Evidence of water flow abounded, with what appeared to be dry river channels, some 600 miles long. As on Mercury and the Moon, asteroid impact had also played an important part in the story, with a crater measuring more than 100 miles in diameter. What did not appear in the photos from Mars was any evidence of an advanced civilization. None of the canals that Percival Lowell thought he had seen. No buildings or satellite dishes. No huts or tilled fields. In fact, no signs of civilization or life.

Intriguing information about the atmosphere arrived in 1974, though, when the Soviet Union's probe *Mars 6* sent back the news of a larger than expected percentage of argon. Because argon is an inert gas, atmospheric scientists surmised that it was left over from an atmosphere—possibly a dense atmosphere, given the large quantities of argon—from which the other gases had since combined with other elements or dissipated into space.

Maybe, given the evidence of running water and a dense atmosphere on Mars sometime in the past, the idea of life on Mars, at least some time ago, was not totally dead.

How could we find out? Two robot spacecraft, named *Viking 1* and *Viking 2*, took off from the launch pad at Cape Kennedy in 1975 partly in the hope of resolving this question. Each spacecraft included an orbiter and a lander, and when the Vikings arrived in Mars orbit in 1976, the landers peeled off to leave the orbiters circling the planet and observing, while they parachuted down to the surface, each in a different spot to gain a rabbit's-eye view of the Martian soil. The lander robots were equipped to scoop up soil from the surface and run a series of tests. One of the biggest questions the Viking science team hoped to answer was the one that had been haunting Earthlings since Lowell's time and before: Is there life on Mars? The disappointing answer came back: No. At least not at the places tested, at least not now.

A mission to study the Martian moon Phobos, launched by the Soviet Union in 1988, met disappointment when the two mission spacecraft failed before reaching Mars, and the same fate met a U.S. project, the Mars Observer, that was scheduled to arrive in 1993.

Many questions about Mars remain unanswered. If Mars did once have a denser atmosphere, allowing water to exist in liquid form on its surface, where and why has all the water gone? Is some of it locked in permafrost just beneath the planet's surface? Exactly how much is stored permanently

The first close-up of Martian soil, taken by the Viking 1 *lander on July 20, 1976. The landing leg is at the right. The dust on the surface of the landing pad at the foot of the lander's leg was stirred up by the landing. The larger chips of rock in this photograph are about 2 inches across.* (NASA Jet Propulsion Laboratory)

frozen as ice at the planet's north pole? We know definitely that some water still exists there. Planetologists have determined that the ice cap that covers the pole is made of carbon dioxide ice, but in summer, when most of the carbon dioxide ice has dissipated, a small water ice cap can be seen. The cap itself sits on top of a curiously thick series of layered shelves that may be composed of dust and water ice.

Perhaps there was once swiftly running water that carved out the gigantic Martian channels. Some scientists, including planetary geologist Michael Carr and exobiologist Chris McKay of NASA Ames (who studies the possibility of life on other planets), think that possibly a very exotic form of simple life may have existed in the "just maybe" more hospitable Martian past and that we may someday discover some trace, some fossil evidence of that past. In the same way that good scientists do, the robot Mars explorers—the Vikings and their predecessors—have answered many questions but have also posed many more.

THE ASTEROIDS

In various orbits in the Solar System, a group of giant, odd-shaped rocks careens through space, most of them orbiting the Sun in a region called the asteroid belt located between the orbits of Mars and Jupiter. You could think of them as "leftovers" from the formation of the Solar System; asteroids are the stuff of which our planets were made, the building blocks of the Solar System, some of which, planetologists think, coalesced billions of years ago to form the planets. Within their composition asteroids doubtless retain a multitude of 4.5-billion-year-old secrets, a wealth of primordial information, if we can only gain access to it.

Smaller groupings of asteroids do lie outside this belt, though, and about 1% of the known asteroids have orbits that cross the orbits of one or more planets. Two of these groups, called Apollo and Aten, for instance, cross the Earth's orbit. Most scientists think that a large one of these Earth-crossing asteroids, as large as six to 10 miles across, collides with Earth an average of once every 50 to 100 million years—the last one probably causing the extinction of the dinosaurs 65 million years ago (see Chapter 5).

There may be as many as 100,000 asteroids that are bright enough for eventual discovery by telescope or spacecraft, but at present only 3,000 have been officially recognized. Of these 3,000 the largest currently known is Ceres, with a diameter of about 633 miles. The smallest probably range down to something less than a mile. In 1991, the spacecraft *Galileo*, headed for Jupiter and its system of moons, viewed the asteroid Gaspra, about 10 miles long and 7 or 8 miles wide, and sent back the first close-up photos of an asteroid.

According to the most recent theory, most asteroids were probably formed when the gravity of the proto-planet Jupiter (early in its formation) prevented the asteroidal material from forming another full-scale planet nearby, leaving the majority of chunks in their present orbit and kicking still others out of the Solar System altogether or into their present planet-crossing paths.

JUPITER THE GIANT

The largest planet in our Solar System, Jupiter intrigues planetary scientists because it is so large and its dynamics are so similar to the Sun's that it has formed what comes close to being its own mini-solar system, with 16 moons orbiting around it. Across the asteroid belt from Mars, Jupiter is the first and largest of the gaseous giant planets. It is so vast, in fact, that it makes up 71% of all mass in the Solar System, aside from the Sun.

During the 1940s and 1950s, astronomers began to put together our modern view of this celestial giant, built on evidence gleaned from Earth-based observations. German-American astronomer Rupert Wildt became the major instigator of the effort to understand the structure of Jupiter, its dynamics and origins. The most striking fact to surface during this pre-space-age period was that the giant planet was more like the Sun than it was like the Earth. The inner planets in the Solar System (those inside the asteroid belt and closest to the Sun) are small rocky bodies with few if any satellites, while planets farther from the Sun are for the most part composed of different stuff altogether. With the exception of little-known Pluto, the outer planets are gas giants, composed primarily of the simplest elements, hydrogen and helium. Most of these outer planets also have an extensive system of satellites. And, as recent explorations have shown, all four of the gas giants—Saturn, Jupiter, Uranus and Neptune—also possess systems of rings circling close around the planet. Taking these major differences between the inner and outer planets into account, modern scientists began to understand Jupiter as a planet with a history much different from Earth's.

Created out of the nebular material that formed the Solar System some 4.5 billion years ago, Jupiter and the other giant planets grew very rapidly and their strong gravity helped them hold onto their original matter. The chemical elements that "built" Jupiter were "sun-like" in nature and remained so. Although Earth and the inner planets were born out of the same nebula, perhaps because they grew more slowly, they couldn't hold onto light gases like hydrogen and helium effectively. So the two groups developed very differently.

But if Jupiter was a body very similar to the Sun and other stars in its composition, why hadn't it burst forth and become a star?

The answer again lies in Jupiter's size. Although large enough to have retained its star-like composition and its family of satellites, it was not large enough to have begun those nuclear reactions deep inside its core that would have triggered a stellar burst. However, Jupiter does put out more energy than it receives from the Sun (unlike the rocky planets)—caused by the tremendous heat generated within by a combination of leftover energy from the planet's formation, its tremendous gravitational contraction and other processes.

The first interplanetary spacecraft to reach Jupiter was *Pioneer 10*, launched in 1972. On June 13, 1983, all its planetary flybys completed, it became the first object created by humans to leave the Solar System. With

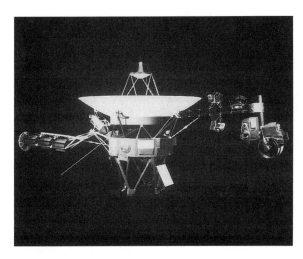

The Voyager spacecraft (NASA Jet Propulsion Laboratory)

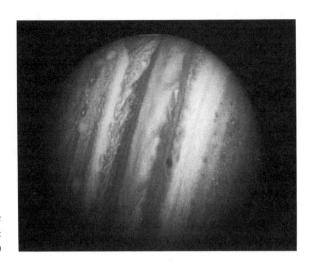

Jupiter at 23.3 million miles (NASA Jet Propulsion Laboratory)

its twin, *Pioneer 11*, it sent back the best data and photographs we had up to that time. But in 1977 the United States launched two Voyager spacecraft that would transform our understanding of the four gas giants of the outer system and provide the most spectacular views of the Solar System ever seen. *Voyager 1* flew by Jupiter in March 1979 and Saturn in November 1980. *Voyager 2* flew by Jupiter in July 1979 and Saturn in August 1981, then caromed past Uranus in January 1986 and Neptune in August 1989. Both Voyagers will exit the outer edges of the solar wind (known as the heliosphere), the far frontier of our solar system, sometime in the 21st century.

For the first time, scientists could see the planet's Great Red Spot close up and observe the violent atmospheric movements taking place there. To their amazement, they discovered that Jupiter has a system of rings, much thinner than Saturn's—only about 0.6 of a mile thick and composed of two parts, one 500 miles wide and the other about 3,200 miles wide.

At Jupiter, the biggest surprises, though, came from the first close-ups of the four "Galilean moons" discovered by Galileo in the 17th century. In some cases one-third to one-half the diameter of Earth, these moons presented a stunning variety: Io, a volcanically active, violently beautiful world with a surface of brilliant reds and yellows punctuated by dozens of jet-black volcanoes; Europa, apparently covered with a thin water ice, with a complex series of criss-crossed linear marks across its surface, possibly huge fractures in the ice; Ganymede, the brightest moon in Jupiter's system, probably half water and half rock, its surface punctuated with dark, heavily cratered areas and many parallel lines of mountains and valleys of more recent origin (possibly indicating tectonic activity similar to Earth's—see

Chapter 5); and Callisto, which, strangely, shows no sign of an active geological history—its thick, icy crust (possibly as much as 150 miles deep) the most heavily cratered in the Solar System. What is the reason for all this diversity?

As planetary scientists continue to study the data from the Pioneer and Voyager missions, perhaps they will find some answers to the many questions Jupiter raises. A new NASA mission called Galileo was launched in 1989 to arrive at Jupiter's system in 1995 and should improve our understanding even further. Unfortunately, the main antenna became stuck in 1990 and some of the data may be lost because NASA is depending solely on the secondary antenna.

SATURN AND ITS ROCKY RINGS

Once seen through the eyepiece of an amateur's telescope, the bright jewel of Saturn and its rings in the night sky can never be forgotten. Galileo, who was the first person to see the strange bulges that turned out to be Saturn's rings, puzzled enormously over what they might be. ". . . The weakness of my understanding, and the fear of being mistaken, have greatly confounded me," he wrote to a friend in 1612. And until the *Pioneer 11* and two Voyager spacecraft sent home close-ups, even our most powerful telescopes did not enable us to unlock the intricacies of their structure.

From Pioneer we learned in 1979 that Saturn is very cold, -279° Fahrenheit, and even colder in the rings, at -328° Fahrenheit, backing up a theory

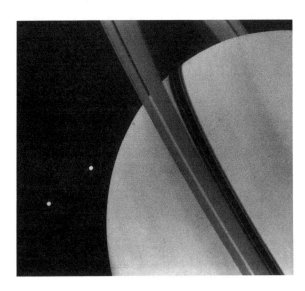

Saturn with its moons Tethys and Dione
(NASA Jet Propulsion Laboratory)

Close-up of Saturn's rings taken November 12, 1980 at a distance of 446,000 miles (NASA Jet Propulsion Laboratory)

that the rings were composed mostly of ice. The pictures Pioneer took of Saturn and its largest moon, Titan, were fuzzy, but they set the stage for the crisper, closer views to come a year later from *Voyager 1* and the following year from *Voyager 2*.

Looking through the Voyagers' remote eyes, scientists saw a planet with much less color than the striking contrasts seen on Jupiter. Since Saturn is colder, it is subject to different dynamics and different chemical reactions. The Voyagers also gave scientists their first close views of turbulence in the belts and zones of Saturn's atmosphere. They measured winds on the planet that reached 1,118 miles per hour, four times faster than those found on Jupiter. And the Voyagers also verified that the giant multi-ringed planet generates nearly twice as much energy as it receives from the Sun, the equivalent of 100 million large power plants.

Saturn's rings, moreover, held thousands of surprises. These regions of orbiting rocky, icy material turned out to be much more complex than anyone had ever imagined. There were not three rings, which is what astronomers thought they saw from Earth, but a complex and constantly changing system made up of tens of thousands of interacting ringlets. The ring system is nearly 249,000 miles across and is composed of millions of tiny particles of ice and snow. Shepherding satellites orbiting close to the rings produced kinks and clumps, even "braids" in the structure. And radial alignments of dust looked like spokes radiating out from the planet through the rings.

Among Saturn's nine known satellites, the most interest settled on Titan, known as the "atmospheric," because observations from Earth had shown that the icy world had an atmosphere of methane and that hydrocarbons might also be present. Larger than the planet Mercury, Titan seemed like a possible candidate for having given a home to some form of life at some time in the distant past. The news from Voyager was even more interesting. The atmosphere of Titan is thicker than Earth's by one and one- half times, and it is mostly nitrogen, with only a small constituent of methane. Before Voyager, Earth was thought to be the only place in the Solar System with an atmosphere primarily of nitrogen. In fact, the nitrogen on Titan registered at levels 10 times those of Earth. Unfortunately, chemical reactions in the thick atmosphere created a dense smog-like condition that made the surface of the satellite itself invisible to Voyager's cameras, so much remains unknown about Titan.

What's going on in that "smog" and under it, on the surface? Since nitrogen is usually a clear gas, and the atmosphere is mostly nitrogen, what's making the haze? The science-fiction mind, of course, can take off from here—but scientists seriously think that the smog could be a kind of organic haze and that the chemical reactions taking place might be similar to reactions that occurred in the Earth's atmosphere billions of years ago. No future mission to explore Titan is currently planned, but the questions the intriguing satellite raises are many.

Overall, Voyager scored a stunning success at Saturn, as *Voyager 2* raced on to see the seventh planet, Uranus, close up for the first time.

MYSTERIOUS URANUS

A far-off greenish ping-pong ball in the night sky, Uranus completely escaped detection until 1781, when it was first sighted by William Herschel. Four times the diameter of the Earth, Uranus orbits far away in a path that is 1,695,700,000 miles from the Sun at its closest. Before Voyager, only five of its moons were known. And, until 1977, little else was known about it except one major oddity: Its axis is "tipped," sloped at 98 degrees. So, unlike the Earth and the other planets, whose equatorial regions point toward the Sun, Uranus spins on its axis almost sideways. It takes Uranus 84 years to complete its orbit around the Sun, and during that time each pole experiences 42 years of light followed by 42 years of darkness.

In 1977, before Voyager's journey, one other fact came to light by accident. A group of astronomers were observing aboard NASA's Kuiper Airborne Observatory, a specially equipped high-altitude aircraft that can fly above much of the distorting haze of Earth's atmosphere. The plan called for observing Uranus as it passed in front of a particular star, a method called occultation frequently used by astronomers to gain more information about one or the other object. To their amazement, the light from the star behind

Uranus dipped slightly before the planet passed in front of it and again afterward. Had they discovered two new satellites on either side of the planet? Further telescope observations confirmed that no, what they had found was that Uranus had rings!

So one of Voyager's prime tasks was to take a close look at the Uranus rings. It turned out that the innermost ring was about 10,000 miles above Uranus's cloud tops and six of the 11 rings (two of which were discovered by Voyager) were only 3 to 6 miles across. The three widest were only about 10 to 30 miles wide. An even bigger surprise was that the rings appeared to consist primarily of large chunks of almost coal-black material, most of them between 3 to 3,000 feet across—much larger than the dust-like particles of Saturn's huge ring system.

Still another curious mystery for the *Voyager 2* scientists was Uranus's strange magnetic field. While most of the other planets in the Solar System are closely aligned with their axis of rotation, Uranus's magnetic field is displaced nearly 55 degrees from its rotation axis. As the planet rotates along its axis, its offset magnetic field wobbles in space. And, the solar wind arriving from the Sun and streaming past the planet draws the far side of the wobbling field into a gently elongated banana shape, an effect never observed anywhere else.

Uranus's atmosphere of hydrogen, nitrogen, carbon and oxygen was found to have a strange sheen rising from the upper atmosphere—possibly ultraviolet light—and the entire planet was shrouded in a smog-like haze, with a surprisingly uniform temperature throughout. Strong, 200-mile-per-

The Voyager 2
*encounter with
Uranus: view of the
planet's rings, with
backlit dust particles*
(NASA Jet Propulsion
Laboratory)

hour winds, twice as strong as Earth's jet stream, moved through the upper clouds. But Voyager could not penetrate the haze.

The moons, however, amazed the science team. Miranda, the innermost moon, gave evidence of tremendous geological activity in its ancient past. It had two different kinds of terrain, one very old and heavily pockmarked with ancient craters, the other young and more complex, obviously the result of major geological changes. Strange racetrack-like patterns and rope-like imprints marred the surface. And, though Miranda is only 300 miles in diameter, a canyon some 50,000 feet deep—10 times deeper than Earth's Grand Canyon—carved a path across its face. Voyager offered quick views as well of the four other largest moons. Ariel had broad, curving valleys and canyons on a young and complex surface. Umbriel was black like Uranus's rings. Titania showed evidence, possibly, of comet impacts in the last 3–4 billion years. And Oberon, like Miranda, showed evidence of enormous fault structures, including tall mountains and craters that appeared to have once been flooded with a dark fluid and then frozen over.

The visit was short, but it left the Voyager mission's science team with much thought-provoking data as Voyager sped away from the Uranian system in January 1986 to head for its next and last stop before speeding toward the edge of the Solar System.

NEPTUNE, OUTER GIANT

Although Neptune, the eighth planet out from the Sun, travels at a distance nearly a billion miles farther away from the Sun than Uranus, it is similar to its distant neighbor in many ways. It is the last, in order of distance from the Sun, of the gas giants, and it appears nearly featureless as seen from Earth-bound telescopes. Its diameter is 3.8 times the size of Earth's, while its mass is 17.2 times greater. Neptune's atmosphere is mostly hydrogen with small amounts of helium and methane, and it is probably the methane that gives the planet its bluish cast. Some scientists also believe that the methane, although slight, may affect the planet's balance of heat by absorbing sunlight. Earth-based measurements indicate that Neptune gives off more heat than it absorbs from the Sun. Some scientists think this excess heat may be caused by a release of energy, as heavy molecules gradually sink toward the planet's core. Although most planetologists think that Neptune has no solid surface, the density of the planet does suggest that it may have a small rocky interior with a mantle of water, methane and ammonia.

But in August 1989 when *Voyager 2* swung past Neptune, it changed forever the formerly faceless image of the great blue planet and its satellites. Before Voyager, only two of Neptune's satellites were known: Triton, which is about the size of Earth's Moon, and Nereid, whose far-flung, eccentric orbit made it impossible for Voyager to photograph it at high resolution. As early as June 1989, two months before arriving at the planet, Voyager had already found another satellite, a dark, icy moon larger than Nereid. Then, close by the planet's ring system, it found a total of five more small moons, similar to the shepherding moons Voyager had found near the rings of both Jupiter and Saturn.

Early in Voyager's approach to the strange, blue ball of gases that is Neptune, the spacecraft discovered an enormous storm system in its atmosphere, the Great Dark Spot, which extends over an area as large as Earth. It is located at about the same latitude as Jupiter's Great Red Spot and its size relative to the planet is comparable. Whipped up by winds as fast as 450 miles per hour, the Great Dark Spot looks like a giant pod exploding at one end as it orbits the planet. It is just one of several enormous storm systems topped by higher-altitude clouds that scoot above them, but scientists are puzzled that Neptune could have such turbulent storm dynamics so far from the energy of the Sun.

Yet another major surprise discovered by Voyager was Neptune's rings, which from Earth had appeared to be incomplete arcs. Neptune's great distance from the Earth, however, had made identification extremely difficult and Voyager established that although the rings are very faint, they do completely encircle the planet as a ring system.

But probably the greatest excitement surrounding Voyager's flyby at Neptune centered on Triton, Neptune's largest moon. This mottled pink body appears to be the coldest place in the Solar System, at about -400° Fahrenheit. (Both Pluto and Charon, which Voyager did not visit, appear from ground-based measurements to be warmer, though much farther away from the Sun.) Triton also appears to have volcanoes of ice, which may even still be active, blasting frozen nitrogen crystals as high as 15 miles into the thin atmosphere.

Neptune was the last stop for *Voyager 2* before heading for the edge of the Solar System and beyond, toward the Milky Way. The small spacecraft and its twin have left behind a great wealth of information and images that continue to stimulate insights as scientists examine and analyze them in hundreds of different ways.

PLUTO, THE FAR TRAVELER

Looking toward the Sun from Pluto, at an average distance of 3.7 billion miles, a future space traveler would see a small, bright spot in the sky no larger than Jupiter appears to us from Earth. The ninth planet from the Sun, Pluto is generally thought to be the last outpost of our Solar System. With its diameter of only around 1,000 miles and a highly eccentric orbit, some scientists have speculated that Pluto may not be a planet at all, but a comet, asteroid or even an escaped "moon" of Neptune.

Pluto wasn't even discovered until 1930, when Clyde W. Tombaugh located it using the telescope at Lowell Observatory in Arizona—aptly enough, since it was Percival Lowell who predicted Pluto's existence based on disturbances in the orbits of Saturn and Uranus. In 1988, scientists traveling in an airplane at 41,000 feet made direct observations of the atmosphere of Pluto, about which planetologists had previously held suspicions, but no proof. There had been considerable controversy over the planet's density. Although we still know little about the planet, it is believed to be composed mostly of water ice with a crust of methane. The discovery of this methane frost on Pluto's surface in 1976 caused planetologists to downgrade their estimate of the planet's size. The methane frost, they reasoned, would make the planet reflect more light from its surface, misleading astronomers into thinking that Pluto was larger than it actually was.

In 1978 Pluto offered yet another surprise when research led to the discovery by James W. Christy and Robert S. Harrington of Charon, Pluto's only moon. With a diameter one-tenth the size of Pluto's, the tiny satellite is the most massive relative to its planet of any in the Solar System.

Charon's discovery further helped give scientists a more reliable estimate of Pluto's mass. As a result, astronomers realized Pluto couldn't be as large as some had speculated; the present estimation of its size is around 1,926 miles in diameter, about one-fourth the size of Earth. In fact, some plane-tologists think that Pluto's small size calls for a demotion from planet to asteroid, but since asteroids don't generally have either moons or atmo-spheres, for now Pluto's status seems safe.

HOW IT ALL BEGAN

The data from all these spacecraft has gone a long way toward filling in the picture of how our Solar System may have formed. Planetary geologists and physicists continue to pore over the millions of photos, images and statistics, and computer modeling helps them test scenarios and measure the results of impacts, temperatures, orbits, angles and velocities. Most agree that the Sun, Earth and eight other planets of our Solar System were formed a little over 4.5 billion years ago by the contraction of a giant cloud (the primeval nebula) of interstellar gas composed mainly of hydrogen, helium and dust. No one is sure what the exact trigger was that began this contraction, but some scientists believe that it may have started when shock waves from a nearby supernova, or stellar explosion, rippled through space, upsetting the delicate balance of the loose cloud's original state. Whatever the cause, once the process started, the contraction continued under the natural and inevitable force of the gravitational attraction within the cloud itself. Contracting and spinning violently, the cloud became disk-shaped and flattened around its outer edges. Meanwhile the huge ball of matter that collected in its center contracted into a heavier and heavier mass to become the developing Sun.

As the cloud continued to spin more and more rapidly, its billions of small particles of dust began to collide with greater violence and frequency, concentrating in the plane of the disk. Gradually these grains built up into larger and larger chunks. And as these "planetesimals," early ancestors of the planets and their satellites, reached a certain size, they no longer depended only upon accidental collision to expand their mass. Instead they began to build up enough mass to use gravity to reach out and attract more and more particles of solid material. These "protoplanets" gradually built up, most of them continuing to rotate and revolve following the direction of the parent nebula from which they were born.

Meanwhile, although the gigantic mass of the slowly developing and contracting Sun had not yet begun to kick on its nuclear furnace, the temperatures in the inner Solar System were high enough to keep volatile substances such as water, methane and ammonia in a gaseous state. Thus, the inner planets formed from such nonvolatile components of the nebula as iron and silicates. But out farther from the embryo Sun, the lower temperatures allowed the volatiles to become incorporated into the formation of the giant outer planets (Jupiter, Saturn, Uranus and Neptune), at the same time allowing these giants to expand by attracting and gathering large quantities of light elements such as hydrogen and helium from the surrounding nebula.

As the Sun continued to collapse and the density and temperature in the inner core began to rise, it approached a critical temperature of around 10 million degrees Kelvin and began to generate energy by the nuclear fusion of hydrogen. Once it was switched on, the Sun also began to generate a solar wind, a flow of electrically charged particles, and like a giant leaf-blower began driving the remaining gas and dust grains out of the Solar System.

Meanwhile, heated by internal radioactivity and the energy generated by the addition of matter, the cores of the protoplanets began to melt, forming the internal structure the planets still have today. Finally, the remaining planetesimals, those too heavy to be swept away by the solar wind, began to blast the developing planets during a great half-billion-year bombardment, the impact scars of which can still be seen on most of the planets today.

WHAT'S IT WORTH?

Human beings have always wanted to know how things work and why, and that's what "pure science" is all about: Finding answers. Raising new questions. Finding new answers and seeing new patterns and the way things fit together. But it has become very expensive to find these answers, and many people rightfully ask, why do we need to know? How can we justify sending spacecraft to other planets when people die on Earth of diseases that can be cured and of hunger that's unnecessary?

Perhaps the best answer is that throughout the short span of time humans have existed, compared to all species that have ever lived on Earth, what measure of success we have achieved at extending and improving human life has always resulted from our search to know and understand.

In this case, the understanding that comes from comparative study of atmospheres, geology, magnetic dynamics and the rest of the physics and chemistry at play cannot be achieved in any other way. The diversity of types and complexity of facts discovered so far has astounded scientists, raised many questions and challenged countless assumptions. The most direct and important result of planetary exploration is the new and urgent understanding we have gained of the delicate nature of our own planet—the uniqueness of its ecosystem in the Solar System and the warning parallels we can see with the lifeless and inhospitable planets we have explored.

MISSION TO PLANET EARTH

*A*fter exploring rocky furnaces and giant balls of gas that orbit with us around the Sun, our Earth seems all the more a welcome oasis of a planet, as seen from space, swirling like a blue, green and white marble. Water sloshes across its surface, the Sun's benevolent radiance gently warming its atmosphere, which is rich in nitrogen and oxygen and covers everything like a protective and nurturing blanket. This is the only planet

Earthrise from the Moon   (NASA)

NASA's space shuttle has launched numerous Earth resource-imaging satellites, as well as countless others for measuring changes in the atmosphere, oceans and crust of the Earth. (Courtesy NASA)

presently known to support life—and space-age exploration of our Solar System has served up a healthy reminder of the fragility of the intricate systems that support living organisms here. In the last half of the century, as geologists, atmospheric scientists, oceanographers and resource specialists explore the forces at work, they have made use of an ever-expanding range of new tools, from comparative planetology to radiocarbon dating to computer modeling to satellite mapping, tracing the history of our planet and predicting its near and distant future. The combined effort has become a worldwide mission to explore the intricate secrets of our own planet, Earth.

THE VIEW FROM ABOVE

Since the success of the first artificial satellite, *Sputnik I*, in 1957, the Earth itself has been under almost constant observation by satellites orbiting high overhead. Although many of these are used for military information-gathering, commercial purposes and communications, many others have studied the Earth's environment and resources. The first Landsat satellite (*Landsat I*) was launched in 1972 and many others have followed since. Circling the globe in a variety of orbits, satellites have discovered the Earth's Van Allen radiation belts, tracked the movements of fish in the oceans, uncovered ancient lost roads and cities in the deserts, and monitored the growth of vegetation and the spread of pollution. Weather satellites have brought us the reasonably accurate five-day forecast plus invaluable information about the state of our atmosphere, including discovery of a growing hole in the protective ozone layer in the upper stratosphere. Resource-imaging satellites track changes in forests and crops worldwide and locate mineral deposits. And oceanographers have used satellite data to study the dynamics of the ocean and its vast, vital currents.

Literally, the space age has both changed the way humankind lives and caused us to see ourselves in the context of the universe around us. But perhaps the most dramatic changes in our view of the Earth since mid-century resulted from a series of bold, new ideas about the Earth's crust, resulting from the use of new tools and methods.

DRIFTING CONTINENTS

This revolution in our understanding of the Earth's crust occurred in the 1950s and 1960s, when geologists developed the idea that the crust of Earth is broken into a number of large plates that move relative to each other. The roots of this idea went back a century earlier, when New York State geologist James Hall (1811–98) noticed that sediments built up along the mountain belts are at least ten times thicker than in interior regions of the continents. From this observation developed the idea that the crust of a continent began as ancient and folded troughs in the Earth's surface that grew thickened through progressive additions of sediment, then hardened and consolidated.

Between 1908 and 1912, German geologist Alfred Lothar Wegener and others came to a recognition that these continents, over long spans of time, tend to rupture, drift apart and eventually collide. These collisions crumple the sediment buildup in the folded troughs, and that's how mountain belts are formed.

Wegener argued that the way the continental edges fit together like a huge jigsaw puzzle reinforced the concept of continental drift. Further, he pointed out that rock formations along the two sides of the Atlantic—in Brazil and Africa—match each other in age, type and structure. They also contain fossils of the same land creatures, creatures that could not have made the long swim across the ocean. Not everyone was convinced, though—especially geophysicists.

That's where the recent story takes up. English geologist Sir Edward Crisp Bullard (1907–79) ran computer analyses on the way these two continents pieced together and found a perfect fit. Other ocean edges did not show the same kind of clear evidence, though, especially around the Pacific and Indian oceans. Many geologists think that mountain chains are still in the process of creation along the edges of the Pacific Ocean, explaining the existence of parallel ranges, strings of volcanoes and frequent earthquakes in these regions.

Studies of the ocean floor with echo-measuring devices in the 1920s had enabled scientists to map and model the sea floor with new accuracy, and this led to the discovery that the mid-ocean ridge first found by 19th-century cable layers in the Atlantic extended over 40,000 miles, nearly around the world. But after World War II, marine geophysicists adapted the military's airborne magnetometer to measure local variations in magnetic intensity and orientation beneath the ocean. This led American geophysicist James Ransom Heirtzler (1925–) to see, from crossing back and forth across the ridge, that the two sides were mirror images of each other.

Further studies, using radioisotope dating of basaltic rocks at the top of the ridge and sediments at intervals down the sides, revealed that the top of the ridge was extremely young in terms of the Earth's history—only about 1 million years old—and the farther away from the ridge on either side, the older the crustal rocks were found to be and the thicker the layer of sediment. The conclusion reached by geophysicists was that the North Atlantic ridge is the spot where new ocean crust is being created, carried up by convection currents as hot lava and immediately cooled on contact with the deep ocean waters. The crustal segments on either side of the ridge move continually apart, about 0.4 inch a year in the North Atlantic, and almost 2 inches annually in the Pacific. These relatively slow rates of movement, powered by thermal convection currents that come from deep in the Earth's mantle, have caused what we call continental drift. This same great gash in the ocean floor connects with the Great Rift Valley that cuts across eastern Africa, where the shifting African and Arabian plates have collided and pulled apart, creating a "floor" bracketed by parallel geological faults. This is a land where intense volcanic activity has played havoc, while upward-shifting segments of rock create jutting shoulders and sinking areas of the floor create great depressions that have filled with water to become the famous lakes of Africa, such as Lake Victoria (the source of the Nile) and Lake Turkana. (These areas have also become some of the richest sources of fossils of ancient humans and human-like creatures ever found, which we'll discuss further in Chapter 8.)

Scientists at Columbia University completed detailed maps of the ocean floor in the 1960s, when Harry Hess came up with the idea that a new ocean crust is being formed at the rift, while the old crust sinks into deep trenches in the ocean floor. (One such trench is located off the Philippine Islands in the Pacific.) Known as sea floor spreading, Hess's theory was confirmed by measuring the ages of rocks dredged from the ocean bottom.

Geologists have synthesized Hess's ideas about sea floor spreading and theories about continental drift into a single theory, known as integrated plate tectonics (*tectonics* meaning movement of the Earth's crust). The crust of the Earth, according to plate tectonics, is broken into several large plates, some of them consisting of crust that's entirely under water, while others have portions of continental crust. Geologists have found, through analysis of seismic (earthquake) data, that a slow-moving, somewhat fluid layer lies 30 to 80 miles beneath the crust, where its shifts contribute to these movements. The plates move, sometimes bumping into each other, causing the formation of mountains, volcanoes and fault lines along which earthquakes occur. One of the plates, for example, contains most of the crust lying beneath the Pacific Ocean; another contains the North American continent and the western half of the crust lying under the Atlantic.

Plate tectonics explains such diverse geological phenomena as the existence of mountains (pushed up by plates overlapping or pushing together), the location of volcanoes and earthquakes (caused by tension between plates), and the formation of trenches and rifts on the floor of the ocean (caused by plates pulling apart from each other).

Although some geologists did not accept the theory at first, by the late 1980s there was firm evidence that plates were moving in the predicted

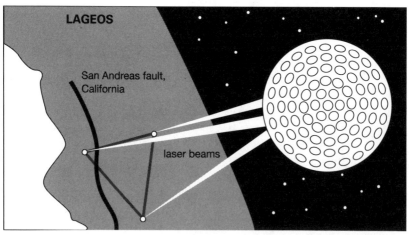

NASA's LAGEOS (Laser Geodynamics Satellite) was launched in 1976 to help predict future earthquakes in the heavily populated San Francisco Bay area. The satellite measures the direction and shifts of the Pacific and North American plates by communicating with three ground stations on both sides of the San Andreas fault.

patterns. The theory has proved successful in predicting many features of the Earth's surface, and in the 1980s, the use of satellites such as LAGEOS and laser measurements enabled scientists to measure the slow-motion movement of the plates at about one inch a year.

DIRTY DEATH OF THE DINOSAURS

Since the 18th century geologists have engaged in heated controversies over the history of the Earth and how it has developed over the millennia. Some highly respected 18th- and 19th-century geologists and comparative zoologists, including biologists Georges Cuvier and Charles Bonnet, thought the Earth's history must have included periodic catastrophes. But they had little evidence to reinforce their claims, and their point of view was overridden by another theory, known as uniformitarianism, which was later replaced by gradualism. Supported by the work of James Hutton and the well-considered theories and writing of Charles Lyell and Charles Darwin, gradualism held that the geologic processes of Earth and the evolution of its living organisms had taken place over long expanses of time, without sudden changes or breaks with the past. Where fossils and geological strata failed to back up this premise, surely a link was missing from the record.

So when Niles Eldredge and Stephen Jay Gould came up with the idea of "punctuated equilibria" in 1972, they knew it was bound to be a controversial idea. In fact, it inspired the decade's most passionate debate over evolution.

Eldredge, who became chairman and curator of the Department of Invertebrates (creatures without backbones) of the American Museum of Natural History in the 1980s, had made a study of trilobite fossils throughout the northeastern United States. Like shrimp and crabs, trilobites—now long extinct—wore their skeletons on the outside and shed them as they grew. So, conceivably, you could find as many as 20 fossilized trilobite skeletons shed by one trilobite over its lifetime, making trilobite fossils a lot easier to find than, say, dinosaur fossils. The tall, thin, bearded young man in glasses scoured upstate New York and Ohio, as well as an area along the Ausable River in Ontario. He found good examples, many of which were around 350 million years old. But he found little evidence that any significant change had taken place in trilobites during long periods of time, as measured by where he found them in the geological strata. Because trilobites have more visible fossil details than most invertebrates—eyes, tail and body ridges—Eldredge was able to make detailed comparisons, using a microscope to measure the distance between the eyes, the height of the eyes and the length of the tail. He compared his findings with similar fossils from Germany and the northern areas of Africa.

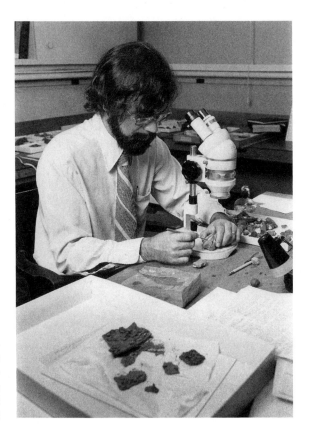

Niles Eldredge (Neg. No. 65849–15A, Courtesy Department of Library Services, American Museum of Natural History)

Then he noticed, as T.H. Huxley and a few others had in the previous century, that the fossil record seemed to show "bursts" of change, or diversification into new species, punctuating long periods of stability. When Huxley had brought this idea up to his friend Charles Darwin, the author of evolutionary theory, Darwin, who saw evolution as a gradual, steady process, had replied that the fossil record was too sketchy to support speculation. But Eldredge's work on trilobites was extensive, and his discovery of relatively short periods of rapid evolution supported Huxley's 100-year-old arguments.

When Eldredge consulted his colleague, Stephen Jay Gould, a professor of paleontology at Harvard University and curator of its Museum of Comparative Zoology, he found enthusiastic agreement. Gould had investigated variation and evolution in Bahamian land snails, which had been around as long as trilobites, and he supported Eldredge's idea that it was time to acknowledge the tale of episodic patterns told by the rocks and fossils. The two of them developed the idea together and published a joint paper describing the evolutionary pattern that Gould dubbed "punctuated equilibria" (often known by the nickname "punk eek"). Gradualism, they argued, was never supported by the evidence offered by fossils and rocks. Instead, change took place in relatively short time frames (lasting perhaps as long as 100,000 years—but that is, as Gould says, a blink of an eye in geological time). And change is interspersed with extended periods of changelessness. Not everyone agreed with this revolutionary idea, but it set the stage for a chance geological discovery published in 1980 by geophysicist Walter Alvarez that suddenly brought catastrophism back to the forefront of scientific thought.

Alvarez was working at a site in Italy, investigating ancient layers of sedimentation to test the rates at which they were laid down. He asked his father, Luis Walter Alvarez, a Nobel laureate in physics, to help analyze some of the clay cores for the presence of certain metals, including iridium. The senior Alvarez had colleagues at the Lawrence Berkeley Laboratory

Stephen Jay Gould
(Harvard University News Office)

who had access to the equipment necessary to check the clay for heavy metals, and he persuaded them to run some tests of the Italian clay. To everyone's complete surprise, one clay layer showed levels 25 times higher than those just above and below it. It so happened that the high-iridium layer was the very layer that had been laid down 65 million years ago, at the end of the Cretaceous age, just before the beginning of the Tertiary age, at what is known as the Cretaceous-Tertiary boundary, often referred to by geologists as the K-T boundary. The question was, why? What's more, when samples of the K-T boundary from all over the world were examined, they showed the same high levels of iridium!

Iridium is a rare element on Earth, but it is contained in abundance in asteroids and comets. So the Alvarezes and their colleagues came up with a highly controversial idea: that the iridium deposits were caused when a large body crashed into the Earth—an asteroid or comet perhaps as large as six or more miles across. Such an impact, they reasoned, could explain not only the high content of iridium and other unusual metals they found, but also the "Great Dying" that took place at the end of the Cretaceous age—the extinction of the dinosaurs and most other life that had dominated the Triassic, Cretaceous and Jurassic ages. When the asteroid hit, they figured, it would have sent huge clouds of dust laden with iridium and other metals common in asteroids billowing up into the atmosphere, blotting out the radiation from the Sun for as long as five years. The lack of sunlight would have cooled the Earth and stopped photosynthesis and killed most plant life on Earth, which in turn would have brought an end to most animal life—some estimates run as high as 75% or more. Eventually the iridium filtered out of the atmosphere to form the layer of rock at the K-T boundary.

The idea caught on in the popular press, and *Time* magazine carried a cover story headlined: DID COMETS KILL THE DINOSAURS? Here was an explanation for the extinction of dinosaurs that had actual evidence to back it up. But the theory was both bold and controversial, and not everyone agreed.

Yet, some took the idea even farther. The well-known mass extinction at the K-T boundary was not the only one in Earth's history. When two investigators, David M. Raup and J. John Sepkoski, Jr., looked at other mass extinctions of the type that occurred at the K-T boundary, they noticed a regular pattern, with similar extinctions occurring about once every 26 to 28 million years. They couldn't think of any factor on Earth that would be likely to cause that kind of cyclical "great dyings." But what about factors from beyond the Earth? The most prominent idea was that something periodically disturbs the cloud of comets—known as the Oort cloud—located at the edge of the Solar System, and when this happens, about 1 billion of the comets fall toward the Sun, and a few of them would almost certainly strike the Earth, causing such catastrophes as mass extinctions.

What could cause such a violent disturbance of the Solar System's equilibrium? One theory centers around the idea that our Sun has a twin, a companion star called Nemesis, that periodically comes close enough to the Oort Cloud at the edge of our little system of nine planets to throw comets or asteroids into a tailspin. The search for Nemesis continues, but the mystery is still open for new ideas and investigation.

Although much of the new catastrophism is still in dispute, the idea that started it all, the impact of an object at the K-T time boundary, has gradually gained acceptance by most geologists and paleontologists. Whether or not the rest of the theory holds, whether or not Nemesis exists, apparently something did strike the Earth at the point in the Earth's history that we now see as the K-T time boundary and caused the mass extinctions. The exact reason is still open to investigation.

Investigations of these ideas, including computer modeling of atmospheric reactions to such a cataclysmic event, also spurred exploration, initiated by astronomer Carl Sagan and biologist Paul Ehrlich, of what would happen to the Earth in the event of the most dread of all human-created catastrophes, nuclear war. And so, the concept of nuclear winter became another spinoff of the Alvarezes' theory about the dirty death of the dinosaurs caused by cooling of the Earth by soot and dust clouds in the sky.

HOLE IN THE OZONE

The last half of the 20th century has seen a dramatic change in our view of Earth as bountiful and indestructible—a change brought about in part by the perspective we've gained by looking at Earth from space. One observation is the disappearance of large quantities of ozone from the stratosphere.

Ozone has been an essential part of Earth's ecology for a billion years or more. Early in the planet's history, once plants began to contribute oxygen to the atmosphere, ozone also began to form through the interaction of solar energy and oxygen. As a result, an ozone (O_3) layer formed—a vast shield centered about 20 miles above the planet that protects terrestrial life from ultraviolet radiation that can be harmful, even lethal. Because ozone has begun to disappear, without its protection humans face increases in skin cancer, and all life-forms may be seriously affected. In addition, due to related chemical reactions, the planet faces a general warming trend that may result in the melting of polar ice caps and widespread changes in agriculture and the balance of Earth's ecology.

Scientists first began to notice that ozone was disappearing from Earth's upper atmosphere in the 1970s, but in 1985 the drama came to a sudden and unexpected head. From data reported by the Nimbus 7 weather satellite, scientists discovered that a gap in the ozone layer has developed high in the skies over Antarctica during the southern hemisphere's spring, each September and October, since 1973.

In 1986, a similar, though smaller, hole was discovered over the Arctic by balloons first launched from Alert, a Canadian town near the North Pole. The scientists for Canada's Department of Environment whose instruments were carried by the balloons found a vast crater above the North Pole, a sort of "drain hole" for ozone of the Northern Hemisphere.

As far back as 1974, F. Sherwood Rowland and Mario Molina warned that chlorofluorocarbons (also called freons or CFCs), commonly used as spray propellants and in refrigeration and styrofoam packaging, might cause a depletion of the ozone layer. By mid-1988, it became clear from investigations that one of the major culprits responsible for the disappearance of ozone in the upper atmosphere was definitely CFCs. In a complicated chemical process, when these chemicals reach the upper atmosphere their components combine with ozone to form other substances. Hence the disappearance of ozone.

In the 1990s, the use of CFCs has finally dropped off. Following worldwide environmental pacts involving most of the nations of the world, several fast food chains have yielded to pressure to eliminate styrofoam packaging, fewer CFCs are used in sprays, and refrigeration is coming into use that does not use the "killer" types of freon. But some scientists urge continued and immediate action. As atmospheric chemist Joe Pinto puts it, "We can't go up there with vacuum cleaners, or pump up more ozone."

EARTH'S GREENHOUSE EFFECT

One of the most thought-provoking products of the new, space-age picture we have of Venus—once thought to be a garden of Eden—is our recognition that Venus doubtless did, like Earth, at one time have great oceans of water. Now those oceans are long-since boiled away by the searing heat of Venus's oven-like atmosphere, and Venus has become a blast furnace in which no creature could possibly live. Could Earth's oceans and atmosphere travel the same route?

On Earth, growing levels of carbon dioxide wastes spewing into the atmosphere from automobiles and factories, combined with CFCs that have damaged the protective ozone layer in the atmosphere, have caused a noticeable overall rise in temperature. Wholesale destruction of rain forests has also disturbed the balance of carbon dioxide and oxygen in the atmosphere. (Botanist Wayte Thomas of the New York Botanical Garden estimated in 1993 that humans have chopped down all but 2% of the Atlantic coastal forest of Brazil since the Portuguese explorers first landed there some 500 years ago.)

Some atmospheric scientists have estimated that, if we continue to expel carbon dioxide into the atmosphere at the present rate, the atmospheric carbon dioxide level will double by sometime in the middle of the next century. Recent computer models indicate that in about 140 years the current levels will quadruple, producing a weakening of the conveyor-belt system of ocean currents that transports heat around the globe. This would create major changes in the ocean circulation, reducing the mixing between the deep ocean and surface waters and limiting the exchange in which nutrients from the deep ocean are brought to the surface and oxygen from the surface is carried to the deep regions. The atmosphere would gradually warm more and more to a temperature equal to what the Earth experienced during the Cretaceous period, when dinosaurs lived and both polar ice caps melted, 65 million years ago. The process, they add, would not be reversible at that level of increase, although if the carbon dioxide level only doubles, the ocean currents will gradually recover over the following centuries.

Although some disagreement exists about these statistics, the message seems clear: Recognize and change the effects we produce on our atmosphere, or suffer enormous consequences in our environment in the coming century.

With a growing recognition worldwide of the fragility of our planet, a new call has gone out to the international community to negotiate the successful management of planet Earth. The lessons learned from the Alvarezes, Sagan and Ehrlich tell us what happens when the Earth's atmosphere becomes dirtied, by whatever means. A look at the conditions on Venus sends up a warning flare about the dangers of a runaway greenhouse effect. Mounting incidence of skin cancer, especially in Australia, signals the urgency of controlling the causes of the hole in the ozone layer. And the fragility of species and their habitats pleads for thoughtful reflection about the natural surroundings we live in and the balance our planet requires. Whether we are capable of becoming good managers, time will tell. But one thing we can be sure of: We can only manage our environment well if we can find out and understand all the ways in which the parts interplay with each other—from the most fundamental and basic constituents. That theme has marked the efforts of physical scientists in the last half-century, and it has infused the work of the life scientists as well.

PART TWO

THE LIFE SCIENCES
FROM 1946 TO THE 1990s

C H A P T E R 6

THE ARCHITECTS OF LIFE: PROTEINS, DNA AND RNA

*B*efore the dawn of the 20th century, the central role played by one complicated molecule in the shape of all living things was one of nature's best-kept secrets. Who could imagine that a molecule known as deoxyribonucleic acid (DNA) was the great architect of life? Or that ribonucleic acid might act as its messenger? The detective route to the discovery was twisted and steep. For one thing, the discovery of these roles required progress in three separate areas: cytology (the study of cells through a microscope), genetics and chemistry.

Like the particle physicists in the physical sciences, life scientists came into the second half of the 20th century with a quest to explore the smallest, most basic building blocks in their field, in this case the essence of living things—proteins, DNA and RNA—at the most fundamental level. Out of this quest a new field was born, a joining of biochemistry and physics in the new molecular biology. It is a field that few a hundred years before could even have imagined—an examination of processes of life at the molecular level that no one had known existed before the beginning of the century, when Gregor Mendel's work in genetics was rediscovered and biologists began to examine the role of chromosomes in heredity.

Although Friedrich Miescher had observed the presence of nucleic acids in cell nuclei in 1869 and in the 1880s Walther Flemming had discovered chromosomes (long, thin structures he spied during cell division), no one knew there was a connection with heredity. Not until Thomas Hunt Morgan began breeding experiments with fruit flies in 1907 (skeptically, at first) did the study of heredity and its tiny, powerful mechanisms begin to take off. By 1911 his laboratory at Columbia University succeeded in showing that the chromosome carried the agents of heredity.

Meanwhile, in chemistry, in 1909 Phoebus Aaron Theodor Levene became first to find that nucleic acids contain a sugar, ribose. And 20 years later he found that other nucleic acids contain another sugar, deoxyribose, establishing that there are two types of nucleic acids, ribonucleic acid (RNA) and deoxyribonucleic acid (DNA). That started the ball rolling on exploring the chemical nature of these substances.

Yet, no one suspected that DNA was connected with heredity. For, while chromosomes contained DNA, they also contained proteins, which seemed much more complex. So proteins seemed like the best candidate for the job of carrying hereditary material—that is, until 1944, when Oswald Avery and a team of other researchers discovered that DNA, not proteins, contain the genetic material of life.

But by 1946, it had become clear that all life-forms employ two categories of chemicals, one that stores information and a second that acts, based upon that information, to duplicate the organism. It had even become clear that enzymes carried out the instructions and DNA held the blueprints, which it also passed on in a nearly exact duplicate to the next generation. What no one yet knew was what structure DNA could possibly have that could enable all this to take place.

THE DOUBLE HELIX

Jim Watson was a tall, lanky young man, who was bright and ambitious from the time he was a kid. In fact, he was on a "whiz kid" radio program called the Quiz Kids when he was 12. Watson graduated from high school at 15 and had two bachelor's degrees (in philosophy and science) from the University of Chicago within four years. From his teen years, he had decided that he would do something that would make him "famous in science." As a stroke of luck, he attended graduate school at Indiana University, where he studied under the distinguished geneticist Hermann J. Muller and worked with Salvatore Luria and Max Delbrück, who had become experts in the study of bacteriophages (viruses that attacked bacteria). After Indiana, Watson went to Copenhagen for postgraduate studies. By sheer chance, he met Maurice Wilkins, who was working on X-ray crystallography of DNA at King's College in London. The idea of doing crystallography—a method of chemical and physical analysis—on DNA, an organic substance, was startling and attracted Watson's attention. Just a few days later, the word was out that at Caltech Linus Pauling, who was the king of chemistry in those days, had worked out a three-dimensional model of the structure of proteins: a helix. The basic shape of a helix is like a spring or the spiral on a spiral notebook (although Pauling's model didn't look much like that).

Linus Pauling
(California Institute of
Technology)

That's when James Watson decided to go to London to learn more about DNA. He wound up, instead, looking for a position at the Cavendish Laboratory at Cambridge, and that's where he met physicist Francis Crick. Crick, who was 12 years older than Watson, was working for Max Perutz on the use of X-ray crystallography to determine the structure of hemoglobin, and his physics background brought new insights to this area.

Watson immediately liked the way Crick thought, and the two men seemed to have an amazing rapport. When they talked, they finished each others' sentences, and they both found the same things fascinating. Through a bit of finagling (Watson's fellowship was supposed to be for study in Copenhagen), Watson went to work with Crick in 1951. They were not technically supposed to work on the structure of DNA, however. That was supposed to be Wilkins's territory, and the Cavendish Laboratory did not want to tread on his toes. So they did their work on the side, on their own time.

102

A few things were already known about DNA. From X-ray crystallography photos that Wilkins's colleague, Rosalind Franklin, had taken, it looked as if DNA formed a helix, too, as proteins did. Also, it was known that DNA consisted of a long chain of nucleotides, and that the chain contained alternating sugar and phosphate groups. A nitrogen base extended off each sugar. Wilkins's work had also shown that (here was a surprise) the molecule was constant in width throughout its entire length.

What Crick and Watson set themselves to find out was how the component atoms were arranged so that they would give a regular structure to the molecule, allow it to be chemically stable and permit it to copy itself faithfully. That is, how could all that fit together? How many helixes were there? Exactly what arrangement did the nitrogen bases have? Did they stick out, making the spiral bristle?

Watson attended a lecture given by Franklin, in which she discussed her preliminary results. Franklin, who had studied X-ray diffraction techniques in Paris, worked with clarity and precision, testing to compare the effect of varying degrees of humidity on her results. She saw that her photographs consistently showed that the molecule had a helical form, but she wanted to test more thoroughly before assuming that the molecule took a helical shape under all conditions. And, even though Franklin produced ever clearer photographs showing the diffractions, the details of the complex structure

Francis Crick (left) and James Watson in 1953 enjoying morning coffee in the office they shared at the Cavendish Laboratory (Cold Spring Harbor Laboratory Archives)

Rosalind Franklin
(Cold Spring Harbor
Laboratory Archives)

remained hard to detect. However, she did describe in detail what she saw and mentioned that she believed that the sugar and phosphate groups, the backbone of the helix, were on the outside, with the nitrogen bases tucked inside. All this, she said, was very speculative. Watson listened attentively but didn't take notes, relying on his usually excellent memory.

Back in Cambridge, with the data Watson brought back in his head from Franklin's lecture, he and Crick were optimistic that they were not far from building a model. How many possible ways could the thing go together? They started building. But Watson, for once, had not remembered the data correctly. For one thing, he didn't remember the figure Franklin had given for the amount of water attached to DNA—about eight molecules of water for each nucleotide in DNA, according to Franklin's estimate. Watson thought she had said about eight water molecules to each segment of the DNA molecule—a far lower amount of water. What's more, he didn't remember Franklin's remark about the position of the nitrogen bases. He and Crick came up with a model composed of three polynucleotide chains with the sugar-phosphate backbone on the inside and the nitrogen bases on the outside. The arrangement corresponded with the data that Watson had brought back from London, and they were confident that they had solved the problem—just 24 hours after they started working on it.

The next day, they invited Wilkins and Franklin, along with a few other colleagues, to come take a look at their model. Franklin's remarks were deflating. Franklin immediately saw the error caused by using the incorrect data and was blunt about it. A molecule designed the way they had built it could not hold a fraction of the water it actually held.

Watson and Crick were thoroughly demoralized and embarrassed.

But they couldn't stop thinking about DNA. And they kept hearing news of Linus Pauling. They knew that Pauling was probably working on the same project, but they also heard enough to know that he was probably on the wrong track.

Several major problems still remained unresolved: How many helixes were wound together in a molecule? Were the nitrogen bases on the inside or the outside? Rosalind Franklin had said she thought they were inside, but if so, how were they arranged? Today a researcher would punch some equations into a personal computer and come up with some what-if models. In the 1950s it was not so easy. So Crick asked a friend, John Griffith, who was a mathematician, to work out how many ways four nitrogen bases might be attracted to each other. Griffith found that electrical forces dictated just two combinations: adenine to thymine and cytosine to guanine.

The second piece of the base-pairing puzzle emerged from a casual conversation that Crick and Watson had over lunch with biochemist Erwin Chargaff from Columbia University. Chargaff mentioned that the "1:1 ratio," about which he had published three years earlier, might be of interest to anyone looking into the structure of DNA. Watson and Crick, unfortunately, knew nothing at all about the 1:1 ratio, and, embarrassed, had to confess ignorance. Chargaff was referring to his discovery, in testing several different biological substances, that equal amounts (a 1:1 ratio) of adenine and thymine occurred, as well as equal amounts of cytosine and guanine. That is, regardless of the species—whether the DNA was from a fish or mammal or reptile—the ratio would be the same.

Crick didn't take long to recognize that Chargaff's 1:1 ratio, together with Griffith's mathematical calculations, were signposts. The nitrogen bases in DNA must pair with each other in a specific way: adenine with thymine, and cytosine with guanine.

But Crick and Watson were still blocked from making real progress. They were not in touch with the cutting edge data that was being produced in this field—because Rosalind Franklin and Maurice Wilkins, two of the key researchers, were not in communication with them. (In fact, Wilkins and

Franklin had a very poor working relationship with each other and barely spoke, even though they were supposed to be working together.) Crick and Watson were not officially working on DNA, of course, and their approach to the problem—building models and then trying to match them to the data—seemed like putting the cart before the horse to Franklin and many others.

Then bombshell news arrived. Linus Pauling had come up with a structure for DNA. Linus Pauling's son, Peter, was doing research at Cambridge at the time, and had become friendly with Watson, a fellow American. So when Peter received a copy of his father's paper on the structure of DNA in January 1953, he passed it on to his friend. Crick and Watson thumbed nervously through the pages. Had Pauling gotten there first?

But no. Pauling had postulated a three-helix model, with the nitrogen bases on the outside, not the inside. This arrangement, Crick and Watson had become convinced, was not correct. He had also made other errors in his paper that surprised the younger scientists.

They still had a chance. Watson decided to make a move, using the excuse of business in London to make an appointment to meet with Maurice Wilkins. The decision—and a sort of seamy serendipity—would provide the key that the two eager young scientists needed. As it turned out, when Jim Watson arrived at King's College, Wilkins was busy, so Watson stopped by to see Rosalind Franklin first. He showed her Pauling's paper, and Franklin was furious, although the reasons why are unclear.[*] Watson assumed it was because she did not think the DNA molecule could exist in the form of a helix. But it's not likely this was the reason, since Franklin, as will become apparent, already knew that at least one form of DNA had a helical structure. Watson tried to show the parallels between Pauling's model and Crick and Watson's first, failed model. This could have irked her, seeming like an attempt to convince her she had been wrong. And she could have been angry that Watson, who was not supposed to be working on DNA, had a copy of Pauling's paper, while she, who *was* working on DNA, had not been sent one by her colleagues in California.

In any case, Wilkins entered the room just in time, according to Watson, who claims he feared Franklin would have attacked him physically at any moment—a rather ludicrous fear, which the 6-foot young man confided in Wilkins after they left her office. Wilkins, who had gotten off on the wrong foot with Franklin from the time she arrived at King's, seemed to open up at this new understanding that Watson now had, having faced Franklin's ire as Wilkins had done on more than one occasion.

---

[*]Only one version of the story—Watson's—exists as Franklin left no written record of this now classic encounter.

Unfortunately, this example shows how scientific discovery, like every other human endeavor, can be either waylaid or aided by the fits and starts of human jealousies and miscommunication. Now, months after the discovery was made, Wilkins finally told Watson that one of Franklin's most telling breakthroughs was that she had found two forms of DNA, which she called form A and form B. Unlike form A, which she had never thought could be definitively pronounced a helix based on her photographs, this new form B clearly had a helical form.

Watson was excited. Could he see a photograph of form B, he wanted to know. As it turned out, Wilkins had been making copies secretly of all Rosalind Franklin's photos—afraid as he was that when she left in a few months for another institution, she would take with her all the work she had done at King's, leaving him with nothing. No doubt, since he had allowed such bad feelings to develop between them, he didn't have the courage to

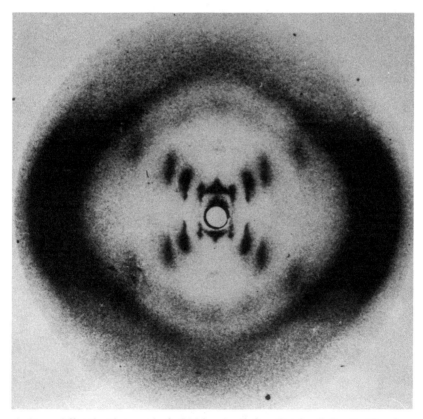

An X-ray diffraction photograph of a DNA molecule, form B, taken by Rosalind Franklin late in 1952 (Cold Spring Harbor Laboratory Archives)

ask for copies. In any case, that's how he happened to have a copy of Franklin's now famous photo No. 51, which clearly showed the structure of form B.

"The instant I saw the picture my mouth fell open and my pulse began to race," Watson later wrote in his autobiographical account of this period, *The Double Helix.* "It was unbelievably simpler than those obtained previously ('A' form). Moreover, the black cross of reflections which dominate the picture could arise only from a helical structure."

Finally, Crick and Watson had the clues they needed. Based on Rosalind Franklin's photo No. 51, they decided to reconsider the double helix. And after five long weeks of trial and error, they arrived at a new model.

The DNA molecule they proposed consisted of two helixes wound around each other, something like a spiral staircase, with the steps made of paired chemical groups of atoms. The nitrogen bases paired with each other, turned inward from two parallel helixes. Then, during replication the two sides of the DNA helix would break apart before the chromosomes split, leaving the nitrogen bases free to pair up again. Each strand of the double helix was a model, or template, for the other. In cell division, each DNA double helix would separate into two strands, and each strand would build up its complementary strand on itself. By matching up the nitrogen bases the same way they were before (no other way was possible), the DNA molecule could replicate accurately.

Watson and Crick admire their model of the DNA molecule
(Cold Spring Harbor Laboratory Archives)

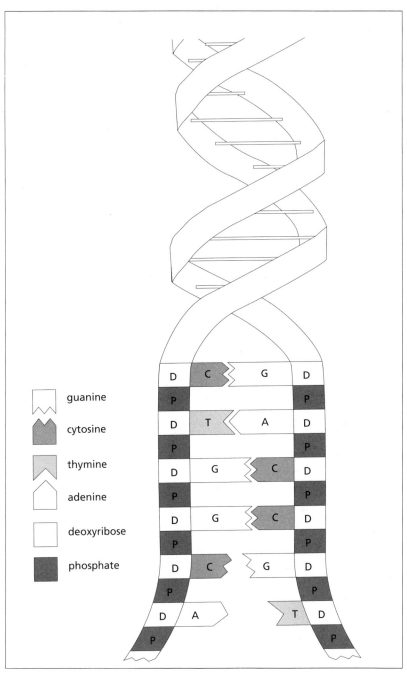

The structure of DNA

The Watson-Crick model of DNA made so much sense that it was accepted almost at once. They had done it! The new model of DNA enabled people to visualize how it could work to guide the construction of other molecules. The basic principle of how one species can reproduce its own kind was at last discovered. It was April 1953.

Watson, Crick and Wilkins received the Nobel prize for physiology or medicine in 1962. (Franklin died of cancer in 1958, before the prize was awarded, and since Nobel prizes are never given posthumously, she did not receive a share of the recognition.) Pauling, who had received a Nobel prize for his work on chemical bonds in 1954, received the Nobel peace prize in 1962 for his opposition to atmospheric nuclear weapons testing (making him the second person in history, after Marie Curie, to win two Nobel prizes).

Few discoveries in science have had both the immediate and far-reaching impact of Crick and Watson's discovery of the structure of DNA. Not only did their double-helix model provide the structure of DNA, but it also predicted a mechanism (using strand separation) by which genetic information could be duplicated faithfully. By 1958, M. Messelson and F. Stahl at Caltech had carried out an experiment showing that DNA does in fact replicate itself by separating the two strands, which serve as templates for creating the complementary sister strands. Moreover, Watson and Crick's model offered the key to understanding how genetic information could be stored in the chemical molecule called DNA.

THE RNA STORY

But Crick and Watson's triumph wasn't the end of the story. Now that the structure was known, new questions arose: How did the DNA convey its instructions to the proteins in the cell? And how were the proteins synthesized? What mechanisms took place?

Using an electron microscope in the 1950s, George Palade initiated a study of microsomes, small bodies in the cytoplasm of a cell. By 1956, he had shown that they were rich in RNA and identified a group of these bodies, which became known as *ribosomes*. The ribosomes, it quickly became apparent, were key to the mechanism of information transfer.

In 1956 Mahlon Bush Hoagland (1921–) discovered relatively small molecules of RNA in the cytoplasm. There seemed to be several different types. Hoagland showed that each type was able to combine with a particular amino acid. The RNA molecule was capable of combining with a particular spot on a ribosome. Amino acids lined up on the other side, so that the RNA molecules hooked up to both and transferred the information from the ribosomes to the proteins (amino acids). Hoagland called the small ribonucleic acids *transfer* RNAs.

But how did the RNA know how to hook up? Transfer RNA was located outside the nucleus in the cytoplasm of the cell, while DNA was deep within the cell's nucleus. A closer look revealed that, in fact, there was also RNA in the nucleus. Two French researchers at the Pasteur Institute in Paris, Jacques-Lucien Monod [moh NOH] (1910–76) and François Jacob [zhah KOB] (1920–), proposed that the DNA molecule transfers the information it holds to an RNA molecule in the nucleus that had used a DNA strand as a model in its formation. These RNA molecules carried the information (message) out into the cytoplasm and were therefore called *messenger* RNA.

Each transfer RNA molecule had a three-nucleotide combination at one end that fit similar combinations at exchange centers on the messenger RNA. So the messenger RNA would settle on the ribosome surface, and the transfer RNA molecules would line up, matching the trinucleotides up to the messenger RNAs on one side and matching up to the amino acids, combined in the correct order, on the other side.

So information was transferred from DNA in the chromosome to the messenger RNA, which then journeyed out of the nucleus to the ribosomes in the cytoplasm to give the information to the transfer RNA molecules. These in turn transferred the information to the amino acids and formed the protein. Three adjacent nucleotides (a trinucleotide) along the DNA molecule plus the messenger RNA molecule plus the transfer RNA molecule made a particular amino acid. The question was: which one?

GENETIC CODE

So, as the 1960s began, the outstanding problem in molecular biology was the genetic code. How could researchers predict which trinucleotide corresponded with the manufacture of which amino acid? Without understanding this process, one couldn't understand how information passed from DNA to proteins.

The stage for exploring the genetic code was set in 1955, by Spanish-born American biochemist Severo Ochoa (1905–), who isolated the enzyme that enabled one strand of DNA to form a second strand from a bacterium. This enzyme, he discovered, could catalyze the formation of RNA-like substances from individual nucleotides. (American biochemist Arthur Kornberg shortly did the same for DNA, and Ochoa and Kornberg received the Nobel prize for physiology or medicine in 1959.)

And that's where American biochemist Warren Nirenberg (1927–) came in. Making use of synthetic RNA to serve in the role of messenger RNA, he began to sleuth out the answer. Nirenberg's breakthrough came in 1961. He formed a synthetic RNA, formulated according to Ochoa's methods, that had only one type of nucleotide, uridylic acid, so that its structure was . . . UUUUUU . . . The only possible nucleotide triplet in it was UUU. So when it formed a protein that contained the amino acid phenylalanine, he knew that he had worked out the first item in his "dictionary." Uridylic acid formed phenylalanine.

Indian-American chemist Har Gobind Khorana, meanwhile, was working along similar lines. He introduced new techniques for comparing DNA of known structure with the RNA it would produce and showing that the separate nucleotide triplets, the "letters" of the code, did not overlap. Independently of Nirenberg, he also worked out almost the entire genetic code. He and Nirenberg shared the 1968 Nobel prize for physiology or medicine with Robert Holley, who also worked in this area.

Khorana later headed a research team that in 1970 succeeded in synthesizing a genelike molecule from scratch. That is, he did not use an already existing gene as a template but began with nucleotides and put them together in the right order, a feat that ultimately would enable researchers to create "designer" genes.

Overall, the first decades after World War II witnessed progress by giant leaps in our understanding of the fundamentals of heredity. *DNA* and *RNA* became household words. And knowledge of the very essence of life seemed just around the corner.

THE ORIGINS AND BORDERLINES OF LIFE: FROM SOUP TO VIRUSES AND DESIGNER GENES

*A*s Crick and Watson and their colleagues delved into the structure of life's architectural processes, other life scientists explored an ever-overlapping pool of issues surrounding the nature of life, how it got started, and how it operated at the simplest levels. Using the new tools of molecular biology and microbiology, investigators came up with more new and astounding insights in these areas.

All of these lines of inquiry led to consequences and breakthroughs on the larger-scale levels, for more complex organisms—most notably for medicine and our understanding of the human body, how it works and how it fits in with the environment around it. This half-century has seen extraordinary interdisciplinary advances that have virtually revolutionized medicine, including CAT scan and MRI imaging, open-heart surgery and organ transplants—most of which are beyond the scope of this book. But the study of viruses and bacteria, as well as the body's immune system, resulted in successful vaccines against polio in the 1950s and made possible an ever-escalating, if still unsuccessful, battle against the AIDS epidemic of the 1980s and 1990s.

THE PRIMORDIAL SOUP

While cosmologists and particle physicists struggled with questions surrounding the birth of the universe and its earliest years, life scientists have grappled with similar questions about living things. Perhaps the most important question of all has been: Where did life come from? The question has intrigued humans for as long as we have any record. The answer is not that easy to figure out—in fact today, some scientists doubt that it can be done at all.

The problem is, as NASA life scientist Sherwood Chang once remarked, "The record is mute. Geologists and atmospheric chemists will tell you there is no evidence one way or the other" about what the Earth was like at that time or how life may have begun. The mystery constitutes one of the greatest scientific detective cases of all time—one that, so far, remains unsolved. With very little hard evidence to go on, dozens of basic questions need to be answered. What were conditions like on Earth before life began? What elements were present? What processes acted on them? What were the original building blocks of life?

For centuries scientists had been trying to figure out whether living things could spring spontaneously out of inorganic substances. Aristotle was sure that they did, and for centuries afterward, scientists tried to prove one way or the other that spontaneous generation could or could not take place. In the 19th century, it seemed, Louis Pasteur had finally provided a definitive answer: It could not. By correcting flaws in previous experiments that had seemed to prove otherwise, Pasteur had shown that a completely sterile, inorganic solution would not produce any signs of life, even with all the needed requirements of life (such as hospitable temperature, presence of oxygen and so on). His trick was to use a special, swan-necked flask that kept out contaminants such as plant or mold spores, while allowing normal atmospheric conditions. But maybe Pasteur's results were not definitive for all time.

An early breakthrough that produced some of the chemicals required for the origin of life was first explored experimentally in the early 1950s by a young graduate student named Stanley Lloyd Miller, who was working on his Ph.D. at the University of Chicago under the guidance of H. C. Urey (1893–1981), a noted American chemist who had received the Nobel prize for chemistry in 1934 for his discovery of heavy hydrogen (deuterium). In recent years, Urey had become interested in geochemistry, the formation of planets and atmospheric conditions in the early years of Earth's history. And he began to wonder about Pasteur's definitive proof that spontaneous generation could not happen. What if, Urey and Miller began thinking, instead of four days, Pasteur had waited the billions of years that Earth had waited for the first life to appear? And what if, instead of the modern mix of nitrogen and oxygen, he had used the primordial atmosphere that existed in the first few billion years of our planet's existence? And what if, instead of a flask full of solution, he had had an ocean full of inorganic molecules?

In a photo taken in 1975, Stanley Lloyd Miller examines the apparatus he used in his 1953 experiment (Courtesy Stanley Lloyd Miller)

For one thing, Urey was reasonably sure that Earth's primordial atmosphere was quite different from today's—most likely made up almost entirely of hydrogen-containing gases such as methane (CH_4), ammonia (NH_3) and water vapor (H_2O). So, under Urey's direction, about the time that Crick and Watson were struggling to work out the arrangement of the double helix, Miller set up a landmark experiment in the history of biology, one aimed at simulating the imagined early-Earth scenario. He postulated a time when great clouds of gases rolled across the turbulent surface of the planet and lightning flashed across the skies. When gas molecules of methane, hydrogen, ammonia and water were jolted into combinations, by ultraviolet radiation from the Sun (since this was before the formation of the ozone layer). And when amino acids and nucleotides formed from the inorganic building-block molecules, and they rained down into the Earth's shallow oceans. Here, Urey and Miller imagined, these pre-organic building blocks bumped into each other and eventually combined into longer, more

115

complex, organic molecules, such as amino acids, proteins and nucleotides. Ultimately, in this imagined sequence of events, these molecules would grow more and more complex until they developed into a nucleic acid capable of replicating itself.

In his small-scale version of this scenario, Miller created an "atmosphere" of hydrogen, with components of ammonia and methane, floating over a flask of carefully sterilized and purified water. Into this primordial "soup" of gas and liquid, he introduced an electrical charge, which he hoped simulated ultraviolet (UV) radiation. Later in Earth's history, plants would begin photosynthesis and produce oxygen, which in turn would contribute to the formation of an ozone layer in the upper stratosphere, protecting the regions below from the Sun's UV rays. But in the beginning, Miller and

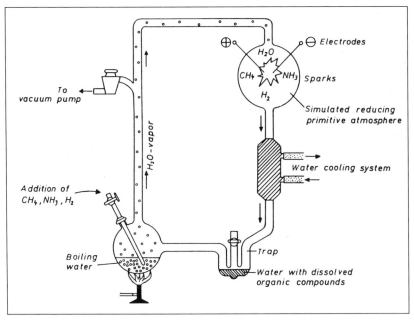

A diagram of Miller's experiment, conducted in 1953 at the University of Chicago. Miller wanted to test whether organic compounds would form if he stimulated the primitive atmospheric conditions that had existed on Earth. Above the flask of water on the right he created an atmosphere of hydrogen (H2), methane (CH4) and ammonia (NH3). The electrical discharges played the part of lightning in the primordial scenario. After operating his experiment for a week, Miller noticed that the water in the flask and in the trap below it had turned orange-red. Tests showed that the water contained amazingly high concentrations of amino acids. The experiment has been repeated successfully many times, producing all 20 protein-forming amino acids. (W. Schwemmler, *Reconstruction of Cell Evolution—A Periodic System*, 1984, with permission of CRC Press, Boca Raton, Florida)

Urey reasoned, there must have been plenty of UV radiation to jump-start the biologic process in this way. Also for now, there would be no free oxygen, but plenty of hydrogen. Miller ran his experiment for a week, and at the end of that time he tested his water solution. He found, to his great excitement, that, in addition to simple substances, he had produced two of the simplest amino acids, plus some indication that a few more complicated ones were in the process of forming.

When he made longer trials, more amino acids formed; other researchers tried the experiment and found they could repeat it. The amazing thing was that the kind of organic molecules that formed in Miller's apparatus were the same kind that are found in living organisms. Miller had not created living creatures, but he had set in motion a process that seemed to be pointing in that direction. Maybe the development of life was not so unusual after all, but a natural consequence of the way the universe is constructed. In the late 1960s, this line of thinking gained even more credibility when more and more complicated molecules were discovered in gas clouds in outer space.

The evidence reinforcing these ideas built up even further when, in 1970, Ceylon-born biochemist Cyril Ponnamperuma found traces of five amino acids in a meteorite that had landed in Australia on September 28, 1969. After careful analysis, he and his team of investigators found glycine, alanine, glutamic acid, valine and proline—the first extraterrestrial constituents of life ever to be found. Ponnamperuma was able to show that the meteorite could not have picked up these amino acids through contamination by its contact with Earth. Clearly they were synthesized by nonchemical processes like those that took place in Miller's experiment.

IN THE BEGINNING . . . CLAY?

But by no means does everyone agree that some version of the primordial soup is the answer to the age-old question. One group of scientists has been working for several years on the idea that life may, in fact, have sprung directly from clay.

Apparently lifeless and inert, clay appears to be the direct opposite of everything we normally associate with living creatures. But the mineral crystals of clay offer some advantages for beginning life that most of us would overlook. For one thing, Sherwood Chang points out that, in the early days of Earth's history (some one and one-half billion years ago or more), the environment would obviously have been dominated by inorganic things. Among those, sea water was certainly present in great abundance. But so was clay.

According to the clay-life scenario, the beginning went something like this: In the early formational days, the basic elements were all present in fair abundance: hydrogen molecules, nitrogen, carbon. Of the ingredients required for organic life, only oxygen was missing. Rocks weathered, shifted and shattered, formed soils and compacted into new sediments, making new rocks and mineral clays. At the molecular level, these clays had (and still have) some special properties. They possessed a highly organized molecular structure, loaded with intrinsic defects that could provide sites for chemical reactions as well as a method for storing and passing on information and energy.

As the environment changed, with alternate wetting and drying, freezing and thawing, and with the movement of wind, water and earth, those clays that could persist and evolve advantageously were the ones to "survive." Graham Cairns-Smith, of the University of Glasgow, Scotland—who originally put forth the theory, delineated in his book, *Genetic Takeover and the Mineral Origins of Life*, published in 1982—thinks these clays developed by adaptive structural evolution or a sort of primordial "natural selection." The clay-life theory contends that these clays may have been primordial life-forms in themselves and, at the same time, a sort of template for living organisms before any living organisms existed. These templates, Cairns-Smith believes, could replicate using a kind of mineral genetic system based on defect structures in crystal formations. And they could adaptively alter these defects (mutations, in a way) and pass them on to daughter generations in classic evolutionary style.

Gradually, organic molecules formed on the mineral molecules, using the clays as a scaffolding. The organic molecules developed an organic genetic system that was more effective and efficient than its mineral predecessor and finally replaced the original mineral formations in a kind of "genetic takeover." In the same way that scaffolding is no longer needed once a cathedral is built and no wooden abacus beads are ever found in electronic calculators, we see no evidence of this "low-tech" life form in today's "higher-tech" organic systems.

Still other theories abound about the origins of life. One of the most favored, supported by biologist Carl Sagan, contends that the carbon required for the formation of organic molecules came from bombardment of our planet by extraterrestrial asteroids containing carbon.

Whichever of these scenarios is closest to the truth, the work done in this field is both exciting and thought-provoking, challenging as it does many long-held assumptions about the line of separation between organic and inorganic and about the nature of organic life and its evolution.

CLONING

While Miller, Urey, Cairns-Smith and others were working on how life may have begun, the work begun in the 19th century on the nature and the

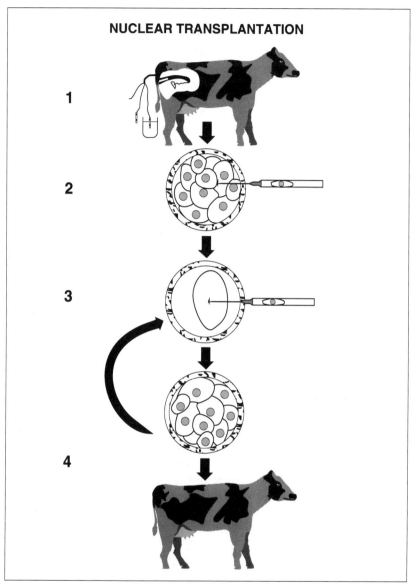

NUCLEAR TRANSPLANTATION

(1) Removal of an embryo from a donor cow (or from in vitro fertilization). (2) Removal of individual cells from the donor embryo. (3) Injection of each cell from the embryo into an egg cell from which the nucleus has been removed. (4) The egg cell grows with its new nucleus to a multicelled embryo stage and then is either re-entered into the cloning process (upward arrow) or placed in the cow's uterus to develop and be born. (Adapted from Office of Technology Assessment, adapted from R. S. Prather and N. L. First, "Cloning Embryos by Nuclear Transfer," *Genetic Engineering in Animals*, W. Hansell and B. J. Weir (eds.), *Journal of Reproduction and Fertility Ltd.*, Cambridge, U.K., 1990, pp. 125–134.)

structure of that basic building-block of nature, the cell, continued. Great strides had been made in the first half of the 20th century, and, thanks to improved staining techniques and microscopes, much of the fine structure of the cell had been identified.

In 1952, however, an extraordinarily delicate operation was successfully carried out by American biologists Robert William Briggs and Thomas J. King in which they removed the nucleus of a cell, which contains all the genetic material of the organism, and substituted it with another. It was the birth of a procedure known as nuclear transplantation.

Fifteen years later, in 1967, British biologist John P. Gurdon successfully cloned a vertebrate, a feat never before accomplished. From a cell in the intestine of a South African clawed tree frog, he removed the nucleus and transferred it to an unfertilized ovum (egg cell) of the same species. From that beginning a new, completely normal, individual developed—a clone of the first.

From Gurdon's breakthrough, coupled with others at the gene and chromosome level in the 1970s, would come new comprehension of how living things function at the most fundamental levels. But in the meantime, a fascinating—and health- threatening—near-creature at the borderline of life posed a challenge to understanding that intrigued researchers for many reasons.

VIRUSES: AT THE THRESHOLD OF LIFE

At the very boundary between living and nonliving, populating a world somehow in-between, is the virus, a bundle of molecules that, like living things, can duplicate itself. But, unlike living things, this parasite can only reproduce inside a living host cell, from which it must borrow essential contributions. The smallest viruses, called viroids, carry only 240 bits of information on their chromosomes, compared to the human body's three billion. But simple as they are, viroids can cause serious plant diseases in the course of their life cycle, which requires duplicating themselves inside the cells of plants. Viruses come in a vast variety of types and different viruses take different things from their hosts, exacting a wide range of tolls.

Viruses and the diseases they cause in humans have played an enormous part in human history, with many viral diseases, such as smallpox, traceable back as many as 2,000 years. But only since the 1880s and 1890s have we known about viruses as the special entities they are. Today, we can confidently say that we have eradicated smallpox, with no cases reported (aside from accidents in laboratories) since 1977. But HIV, human immunodeficiency virus, the cause of the AIDS epidemic of the 1980s and 1990s, is another matter. So far, it is well beyond our ability to control or understand.

Obviously, medical concerns provide an urgent motivation for studying these tiny creatures on the borderline of life.

Viruses, for example, may give us a new understanding of cancer, its causes and, potentially, its prevention. By confusing the host cell—distorting its functions and signals—viruses sometimes can cause cancer. By identifying these distorted functions and signals, we have come to a new, molecular understanding of what happens in cancer, pointing the way to new treatments and approaches. In a revolutionary test now under way in Taiwan, 63,500 newborn babies are receiving vaccinations against infection with the hepatitis B virus. Scientists predict that this preventative step could stop as many as 8,300 cases of liver cancer in these babies, before it ever gets a chance to get started.

But viruses also provide a sample of an organism at work at the very simplest level, and so they give researchers a window for understanding more complex living things.

As German-American microbiologist Max Delbrück (1909–81) wrote in 1949, "Any living cell carries with it the experience of a billion years of experimentation by its ancestors." (With microbiologists Alfred D. Hershey [1908–] and Salvatore E. Luria [1912–91], Delbrück won the 1969 Nobel prize for physiology or medecine for their work on the genetic structure of the virus.) As Delbrück implied, by observing at these levels, we have learned a great deal, and we continue to do so as we ask and answer questions: What are viruses? Are they alive? What do they look like? Are there many different kinds of viruses? Does each virus cause one specific disease? How do viruses cause diseases? How do viruses duplicate themselves? What have we learned about viruses that can be applied to humans?

By the 1930s, scientists had agreed to use the term *virus* to refer to any agent that could pass through a filter that would ordinarily retain bacteria; today it has been refined to refer only to submicroscopic agents (less than 0.3 micron). They all require a host cell in order to duplicate themselves, and this parasite behavior often causes the death or alteration of the host cells—which is why most viruses cause diseases.

Also in the 1930s, two events contributed to an expanding understanding of the virus and its nature. First, Wendell Stanley showed, in 1935, that a virus can take on a crystalline form. And second, German electrical engineer Ernst August Friedrich Ruska (1906–88) introduced a new tool, the electron microscope, which uses electron beams instead of visible light. The instrument's powerful focus permitted the resolution to increase initially by 400 times and was rapidly improved. (Somewhat late, Ruska received the 1986 Nobel prize for physics, for his work.) By the late 1940s and 1950s, the magnifications were improved to 100,000 times and objects as small as .001 micrometer in diameter could be resolved. From 1959 to 1961, Finnish cytologist Alvar P. Wilska and French cytologist Gaston DuPouy worked out ways to set the parts of the virus in relief and use an electron microscope on living viruses. Now electron micrographs enabled scientists to see the microscopic world of viruses, whereas before they could only guess.

Viruses have a nucleic acid core containing genetic information that is protected on the outside by a protein coat and, in some, a fatty envelope around that. The shell is in the shape of a geodesic dome with 20 equilateral triangular faces. Some viruses use DNA to carry their genetic code (as nonviral living organisms do), and others use RNA.

As for the question whether viruses are alive, the debate continues. They share many characteristics with living things, and they use the same genetic code as living things. They correlate with many aspects of living cells—they couldn't succeed to live off the cells if they didn't. They function according to a plan and can program their own replication within the confines of a cell. Like other living things, viruses evolve and respond to changes in their environment. Does this describe the simplest of living things? Or does it describe an extremely complex set of molecules—a handful of chemicals?

Retroviruses

Retroviruses were encountered initially around 1908 in chickens, where they were discovered to cause leukemia, but no one at first suspected their strange life cycle. The retrovirus chromosome, it turns out, consists of RNA, not DNA. Originally known as an RNA virus, the retrovirus produces a DNA copy or provirus, which in turn is transcribed into viral RNA. (This is backwards, hence *retro*.) In 1970 Howard Temin and David Baltimore independently announced the discovery of an enzyme in the virus particle itself—the reverse transcriptase—that copies RNA to DNA (the opposite of the transcription that takes place in most cells). This process was key to the way certain viruses replicate. It also provided a new tool that allowed molecular biologists to copy any RNA species to DNA. This was a key step in the 1970s revolution of gene cloning and genetic engineering. Lessons from the study of retroviruses have also formed a cornerstone of our modern understanding of cancer.

The retrovirus attaches itself to a specific receptor on the surface of the host cell and then penetrates into the cytoplasm of the host cell. There, this sneaky invader shirks off its protective coating, and then, through reverse transcription, turns its RNA into a double strand of DNA. This double-stranded DNA moves into the nucleus and is integrated into the host cell's set of chromosomes (or genome), where the DNA creates a DNA copy (called a provirus), which in turn is transcribed into viral RNA. Now the virus itself is inherited by the offspring of the host cell. When, during the cell's normal replication process, the RNA is transported out of the nucleus and onto the cell ribosomes to be translated into proteins, certain enzymes associate with two copies of the viral RNA in the cytoplasm and move to the plasma membrane surrounding the cell. There, the budding process takes place: The virus particle surrounds itself with an envelope borrowed from the plasma membrane of the infected cell, is released from the surface of the cell, and travels off to look for a new host cell and begin the cycle again.

Some retroviruses don't kill the host cell, which goes on carrying the provirus in its DNA structure and multiplying it as it replicates. Other retroviruses cause changes in the host cell that result in tumors. A third type, of which HIV is a member, kills the host cell by a means that remains unclear.

While other retroviruses have usually not been impossible to control, HIV remains the greatest challenge to face researchers in the last quarter of the century. It has turned out to be a subtle, determined and mean adversary.

Subtle and Mean: The Story of AIDS

In the late 1970s, physicians in New York and San Francisco began to encounter some unusual cases of fungal infections and a rare cancer called Kaposi's sarcoma. By 1980–81, at the Centers for Disease Control (CDC) in Atlanta, Georgia, which distributes experimental drugs for cases of rare diseases, a technician named Sandra Ford began to notice a suddenly high number of cases of *Pneumocystis carinii*, a rare form of pneumonia that usually struck only those with greatly weakened immune systems, most often as a side effect of chemotherapy for cancer. But these cases were otherwise healthy men. So the CDC began looking further and found that some 500 cases of a mysterious disease had been reported—a lethal affliction that knocked out the victim's immune system. About four-fifths of these first known cases were homosexual and bisexual men, and in the beginning the disease was named the Gay-Related Immune Disorder, but the name was soon changed to Acquired Immune Deficiency Syndrome, or AIDS. In the early 1980s, this now widely known disease was relatively obscure. But by the end of 1982, more than 800 cases had been reported in 30 states.

Then in August 1985, the news broke that popular film star Rock Hudson had AIDS, and he died of the disease shortly thereafter. Now AIDS suddenly burst upon general public awareness as a very serious problem with a rapidly increasing number of victims. It also soon became clear that the disease was not confined to the male homosexual population. The disease had been active in the general population in Africa and Asia since the late 1970s, and studies estimate that between 5% and 20% of the entire sexually active adult population in some areas of Africa were infected by 1992. Also by 1992, AIDS had become the leading cause of death in women of reproductive age, between the ages of 20 and 40, in the major cities of North and South

America, Western Europe and areas of Africa south of the Sahara Desert. What's more, studies indicate that between 24% and 33% of the children born to women infected with HIV will develop AIDS.

The questions were many and now, finally, were recognized as urgent. What caused AIDS? How was it transmitted? How could it be cured, or even controlled? By this time it was clear that AIDS was an infectious disease, and it soon became recognized that a virus known as the HIV-1 virus was the probable cause (discovered separately in 1983 by Luc Montagnier at the Pasteur Institute in France and by U.S. National Institutes of Health researcher Robert Gallo). The virus spreads virtually exclusively by infected blood or sperm—although researchers have also found HIV in several other body fluids, including saliva, tears, urine, breast milk, cerebrospinal fluid, and certain cervical and vaginal secretions. Initially, AIDS was sometimes transmitted by blood transfusions and, in some cases, in clotting factors used to treat hemophilia (a hereditary disease typified by excessive bleeding). One of the first defense moves was to begin testing the blood supply and the clotting factors supply for the HIV-1 virus, and, as a result, that source of infection was virtually eliminated by 1985.

In January 1983 came the first reports of cases in the United States among heterosexual victims who were partners of intravenous drug users. The disease can be transmitted by very little infected blood—in fact, health officials discovered that when drug users pass a used needle from one person to another, an infected drop of blood on the needle is all it takes to transmit AIDS. Intravenous drug users rapidly became recognized as a major group at risk for the disease.

AIDS attacks the host's immune system and that's how it kills. The immune system is the part of the body that defends against disease and infection by fighting off viruses, bacteria and other invaders. HIV attaches to a specific receptor on the host cell's surface, climbs inside the cell, replicates there and kills the cell. HIV's favorite receptor is an important member of the human body's immune system known as the CD4 T cell. The CD4 T cells recognize foreign attackers in the bloodstream and help other cells in the system make antibodies. They also help a third type of cell develop into killer T cells. So the loss of CD4 T cells has a devastating effect on two arms of the immune system, affecting both the production of antibodies to fight off attacking viruses and the creation of killer T cells.

Death is not immediate—and in fact seven to nine years or more can pass between infection with the virus and development of AIDS. During this time a victim can be a carrier without knowing it. By the early 1990s, one estimate set the figure between 1 million and 1.5 million people in the United States infected with the virus. Once AIDS develops, the progress of the disease can move slowly, but the end result is always the same: repeated bouts with opportunistic illnesses that take advantage of the weakened immune system and, finally, death. Between 1981 and 1991, some 190,000 cases of AIDS were reported in the United States; 120,000 people had died from it. Worldwide, the World Health Organization reported 203,599 cases of AIDS by the end of 1990, with estimated numbers ranging realistically as high as 600,000. And in East Africa alone, estimates predicted some 400,000 cases by the end of 1993.

What makes the HIV-1 virus uncommonly deadly is the way it disguises itself as part of the body and then disables the T cells, a key part of the immune system. Finding a cure becomes particularly elusive, since the HIV virus readily transforms itself to resist the toxins of any drug. Then the new, mutated virus goes on about destroying and replicating in a new form, unaffected by the drug and unrecognized by the body's immune system.

Several drugs have been found that effectively slow the progress of the disease, including azidothymidine (AZT) and dideoxyinosine (ddI), although they have many toxic side effects. Hopes are constantly raised that a cure may be just around the corner, but the job is tough. The HIV virus changes its form more than 1,000 times faster than most influenza viruses, and trying to develop either a vaccine or a cure to defeat it is, as some scientists put it, "like trying to hit a moving target." In addition, most cures for disease try to bolster the body's own immune system to win the battle. But in this case, the immune system is the very system under attack. One such hope arose, however, in February 1993, when researchers studying HIV in the laboratory found that some combinations of drugs—especially a particular three-way combination—produced a hopeful effect on the virus. Although the virus was able to mutate, as usual, to avoid the toxic effects of the drugs, the new version of the virus was sometimes unable to replicate. This meant that, once the particular virus individual had lived its lifespan, it would die without offspring. Eventually, if this worked over the long term and consistently, the virus would completely die out in its host and the war would be won. But by October it became apparent that this three-fold strategy doesn't always work. Sometimes HIV that is resistant to multiple drugs and can continue to survive and replicate does evolve. In fact, in tests by other researchers, the strategy didn't work; the mutated HIV now seemed able to survive and continue its normal life cycle. So the three-drug attack doesn't necessarily lead to a disabled virus. Neither has any vaccine turned up, despite efforts to find one, and the prospects don't look bright. As one researcher cautions, because of the nature of this virus, a vaccine could backfire, mistaking the immune system itself for the virus it has been set out to destroy. And so, there is still no magic bullet for AIDS.

As a result, AIDS remains a great frustration. In this, the century of science and its triumphs, we have come to expect the kind of security produced by sulfa drugs, penicillin and an army of immunizations—result-

BARBARA McCLINTOCK AND THE CASE
OF THE SHIFTING GENE

Understanding genes has never been easy. When botanist Barbara Mc-Clintock presented a paper at a Cold Spring Harbor Symposium for Quantitative Biology in 1951, few people really understood what she was talking about. A highly independent researcher, she was saying, based on her careful studies of generation upon generation of genes in corn, that some genes shift easily and frequently from place to place on chromosomes.

"Transposition," as McClintock's observations came to be called, clashed with the copious work on the genetics of fruit flies (*Drosophila melanogaster*) that Thomas Hunt Morgan and his team accomplished in the early decades of the century.

When McClintock began her work in this area, in 1944, she had just been elected to the National Academy of Sciences and had taken office as president of the Genetics Society of America, the first woman to hold that position. She recognized that she was cutting against the grain with her theory, but, from the beginning, she sensed that the evidence led in the direction she was testing, even though for two years she was in limbo, not knowing whether the line she was pursuing would pay off or be a blind lead. She later remarked, though, to Evelyn Witkin, a colleague at Cold Spring Harbor Laboratory:

ing in longer and more productive lives than ever in history. Can we find a way to outwit this deadly killer, this tiny adversary with its extraordinary evolutionary ability to adapt? Only time will tell.

Meanwhile, our best defense against AIDS has become education. If we can succeed in preventing HIV from traveling from host to host, eventually it will die out. The key to this plan is to make everyone aware of the high risks attached to sexual relations without protection ("unsafe sex") and intravenous drug use (especially, sharing needles). These are two of the three most likely ways to contract AIDS. (The third method by which the HIV virus is most commonly transmitted is from mother to child during pregnancy.) Once infected with HIV, as far as the research shows, the host ultimately develops AIDS and dies from it. But everyone (except the child born of an infected mother) can protect himself or herself, just by avoiding risky behavior.

For the many thousands of afflicted individuals, the hope for a cure continues. For the many thousands more who may otherwise develop the

It never occurred to me that there was going to be any stumbling block. Not that I had the answer, but [I had] the joy of going at it. When you have that joy, you do the right experiments. You let the material tell you where to go, and it tells you at every step what the next has to be because you're integrating with an overall brand new pattern in mind. You're not following an old one; you are convinced of a new one. And you let everything you do focus on that. You can't help it, because it all integrates. There were no difficulties.

McClintock was far too rigorous a scientist to rely on the reading of tea leaves. But, like most productive scientists, she played hunches—which in reality were based on many years of close observation and a fine ability to read patterns. The phenomenon she found was a complex process of regulation and control that, at the time, was unlike anything yet observed. Prompted by a control mechanism, a chromosome could break, dissociate and rejoin differently—all in systematic, observable ways. She arrived at this conclusion by observing changes in color patterns in kernels of Indian corn and correlating them with changes she observed in the chromosome structure.

Only recently has McClintock's work on transposition become recognized as a fundamental and revolutionary concept of gene functioning. For many years, like Gregor Mendel's, her work on transposition was ignored; she was ahead of her time. However, McClintock turned out to be right—and when other investigators began to find evidence that genes did sometimes move around, they remembered that she had said it first.

In 1983, Barbara McClintock received the Nobel prize for physiology or medicine. She was 81.

disease in coming years, the cry for a vaccine and a cure has not stopped. And the search goes on.

BIRTH OF GENETIC ENGINEERING

While HIV continues to elude understanding, though, spectacular advances in other areas have taken place. At mid-century, the breakthroughs made by Crick and Watson at the molecular level coincided roughly with investigations that other biologists were conducting on a special, fascinating class of viruses that attacked bacteria. Known as bacteriophages ("bacteria eaters"), or just phages, this group of viruses had an unusual characteristic that led ultimately to the discovery of ways to transfer genetic material from one organism to another. The combination of new understanding of these mechanisms and new technology led to one of the most stunning scientific advances of the century: genetic engineering.

Some of the first steps, though, were in the study of bacteria, not their parasites. Joshua Lederberg started down this path in 1952, when he noticed that bacteria exchange genetic material by conjugating, or joining together in pairs, in a process much like sexual intercourse in more complex organisms. Lederberg also observed that there were two groups, which he named M and F. The group he called F all possessed a body he named a *plasmid*, which the F bacterium passed on to the M bacterium. The plasmid, it turned out, consisted of genetic material, as William Hayes discovered the following year. By this time it had been clear for several years that the genetic code was carried by DNA; the plasmid, it seemed, was a ring of DNA, floating separately from the main DNA in the chromosome of the bacterium.

This discovery brought immediate help for solving a growing problem in the medical field. Sulfa drugs and antibiotics developed in the 1930s and 1940s had been in use for several years, so that many bacteria had begun to develop resistance to them—and unstoppable epidemics had begun to threaten again, especially in hospitals. In 1959, a group of Japanese scientists discovered that the gene for resistance was carried by the plasmid, of which one bacterium could have several copies, which were then passed on from one bacterium to another. A few drug-resistant bacteria introduced into a colony resulted in an entire colony of drug-resistant bacteria in short order.

As early as 1946, meanwhile, Max Delbrück and Alfred D. Hershey, working independently on bacteriophages, found that genes from different phages could combine with each other spontaneously. Swiss microbiologist Werner Arber took a look at this strange mutation process in detail, and he made what would turn out to be a stunning discovery. Bacteria had an effective method for fighting off the enemy bacteriophage virus: They used an enzyme, which became known as a "restriction enzyme," to split the phage DNA and restrict the phage's growth. The phage became inactive, and the bacterium went on about its business.

By 1968, Arber had located the restriction enzyme and found that it was designed to locate only those DNA molecules that contained a certain sequence of nucleotides that was characteristic of the enemy bacteriophage.

Arber watched what happened. The split phage genes fought back by recombining after they were split. Once split, he found, the split ends of the DNA are "sticky." That is, if the bacteria's restriction enzyme wasn't present to prevent recombination, different genes that had been split at the same location would recombine when placed together. The birth of recombinant DNA—that is, DNA that has been artificially prepared by combining DNA fragments taken from different species—was close at hand.

On the heels of this discovery, in 1969 Jonathan Beckwith and coworkers succeeded for the first time in isolating a single gene, the bacterial gene for a part of the metabolism of sugar. Now the stage was set for genetic engineering.

GENETIC MARKERS AND THE
HUMAN GENOME

Some diseases are not caused by a parasitic virus or bacterium, but are genetic—that is, passed on from generation to generation. In fact some 3,000 diseases are known to be caused by specific genes. Huntington's disease, caused by a single gene, is one example; cystic fibrosis, caused by a combination of two genes, is another. It would be enormously helpful if we could isolate the specific genes that caused specific diseases, but this turns out to be a tall order, since one human being's characteristics are produced by some 2,000,000,000 genes. (The full complement of genes that produce an individual human is known as the human genome.)

But it isn't impossible. DNA can be divided into lengths, using the restriction enzymes produced by bacteria. And techniques have been developed for sorting these short pieces of DNA by length, followed by a technique known as Southern blotting (named after inventor Edward A. Southern) for picking out particular pieces of DNA from the sorted lengths.

In a family in which some members have inherited a particular disease, if you can compare lengths of DNA from each member of the family, and if you know which members of the family have the disease, then it's often possible to identify a particular piece of DNA that appears similar in the affected family members. This piece of DNA becomes a marker for the gene that causes the disease. Ultimately, scientists can use this marker to find the chromosome (one of 23 pairs) on which the gene is located. In 1983, Kay Davies and Robert Williamson found the first marker for a genetic disease, Duchenne muscular dystrophy, using these methods.

In 1988, the U.S. National Academy of Sciences called for a major national effort to map all the genes in the human genome and the undertaking soon became international in scope. Five years into the project, this effort was already proving to be one of the greatest endeavors in the life sciences in the modern era.

By the early 1970s, American microbiologists Daniel Nathans and Hamilton Smith picked up the baton and began to develop various restriction enzymes that break DNA at specific sites. In 1970, Smith discovered an enzyme that broke a molecule of DNA at one particular site. Nathans worked on this process further and found ways to create tailor-made nucleic acid fragments, of a sort, and study their properties and ability to pass on genetic information. Now investigators were truly on the road to work with recombinant DNA that would make it possible to take nucleic acids apart and put them together in other forms. Smith and Nathans shared the 1978 Nobel prize for physiology or medicine for their landmark discoveries.

In 1973 Stanley H. Cohen and Herbert W. Boyer combined the two techniques—the ability to locate restriction enzymes in plasmids and the isolation of specific genes—to introduce an extraordinary breakthrough that would become known as genetic engineering. They cut a chunk out of a plasmid found in the bacterium *Escherichia coli* and inserted a gene from a different bacterium into the gap. Then they put the plasmid back into *E. coli*, where the bacteria made copies as usual and transferred them to other bacteria. It was an amazing and powerful trick. Other scientists jumped in within months to repeat it with other species, inserting genes from fruit flies and frogs into *E. coli*.

But not everyone thought this was a good idea. In July 1974, Paul Berg and other biologists held a meeting backed by the U.S. National Academy of Sciences to draw up guidelines for careful regulation of genetic engineering. The tension has continued since that time between those who want to explore genetic engineering further and those who worry about undesirable consequences and want to control it.

But by the 1980s, genetic engineers succeeded in producing several proteins needed by those who were sick, such as human growth hormone, human insulin, interleukin 2 and a blood clot dissolver. They also came up with a vaccine for hepatitis B and improved tissue matching capabilities for transplants. Most of these enterprises used tanks for production in carefully controlled environments, and opposition to this type of genetic engineering has lessened. Additionally, genetic engineering has succeeded in locating gene markers for such genetic disorders as Huntington's disease or Duchenne muscular dystrophy (see box page 129).

Genetic engineering enjoys an uneasy peace with both the public and science. Many people hesitate about such experiments as releasing genetically engineered bacteria into the environment, and even an apparently innocent product such as genetically engineered tomatoes seems to have a negative public image. But, responsibly used, genetic engineering can be a powerful tool for managing our lives and our health, and it is, without question, one of the most important discoveries of the century.

CHAPTER 8

WHERE DID HUMANS
COME FROM?
THE SEARCH CONTINUES

The Great Rift Valley that slashes across seven countries from the southern tip of Turkey to Mozambique looks like a rough-edged scar across the arid face of most of eastern Africa. Barren, hot and forbidding, lashed by eroding winds, this badlands region of jutting shoulders of rock and exposed ancient geological layers has become a giant treasure chest for those who seek to know the history of humankind. For here, near several ancient lake beds, skeletons that are hundreds of thousands to millions of years old have been found. Thanks to the shifting along fault lines, these skeletons are lifted up from where they were buried beneath eons of rock and sediment deposits and are exposed by the erosion of constant winds. Eventually, a passing paleontologist, by training always ready to notice the odd fragment or a telling shape or color, picks it up and catalogs it—and the search is on for the rest of the skeleton, for other skeletons, and, possibly, for tools and other evidence of a way of life.

The work in this land is not easy. And the piecing together of the history of hominids—a term used to describe us and our two-legged, upright-walking ancestors—requires keen eyes, patient and careful hands, often several teams of scientists from varied disciplines and, above all, a thick skin and a lot of charm. Paleontology is among the most controversial of sciences. Because paleontologists draw conclusions, necessarily, from fragments of bones and often scanty evidence, fellow scientists tend to look down on their theories, which non-paleontologists think say too much based on too little data. Even, or perhaps especially, among themselves, paleontologists find agreement hard to come by. What's more, their business is the exploration of one of the touchiest, most emotional of subjects: exploring the idea that human beings have descended, at the very least, from ape-like creatures, if not directly from apes.

When Charles Darwin first put forth this idea, the bishop of Worcester's wife is said to have remarked, "Let us hope it is not true, but if it is, let us pray that it will not become generally known." And that is the unqualified opinion still held by many people. As a consequence, funding is hard to come by in this orphan profession, although one institution and one family have gone a long way toward capturing the public interest and enthusiasm: the National Geographic Society and the Leakey family of Kenya.

THE FAMOUS LEAKEY LUCK

Louis Seymour Bazett Leakey (1902–72) was born in Kenya, where his British-born parents had gone as missionaries, and to him, Kenya was always his home. A maverick, highly independent, tall, intuitive man with a twinkling eye, he did more than anyone before him to bring the hunt for the hominid story to public attention in a positive light. He was also enormously controversial. But Leakey and his wife, archeologist and paleoanthropologist Mary Nicol Leakey, together transformed the entire picture of human prehistory.

Mary Leakey (1913–) was the one who actually ran most of the sites, setting up systems that became the standard of the profession, and she made many of the finds. Systematic, careful and attentive to detail, she contributed the perfect complement to Louis's impulsive, swashbuckling, hunch-playing approach. "If we'd been the same kind of person," she has asserted, "we wouldn't have accomplished as much." A trained artist, Mary also illustrated most of Louis's publications.

One morning in 1959, Mary was walking the dogs at a site in the Olduvai Gorge in northern Tanzania, where she and Louis had been working (Louis was back in camp with a fever). Her ever-watchful, trained eyes lit on something striking, in an exposed cross-section about 20 feet down from the top layer. A closer look revealed a bit of exposed bone, which at first didn't appear to be hominid. But when she brushed away the soil, and could see the teeth, she became excited. "I was tremendously excited by my discovery and quickly went back to camp to fetch Louis," she later wrote. Louis was not quite as excited as Mary. To him, it looked like an *Australopithecus* skull, and he was looking for fossils in the *Homo* species.

Australopithecus was a name first coined by Raymond Dart in 1924 to describe the fossils he had found in South Africa that became known as the "Taung baby." *Australopithecus* means "southern ape," but most researchers had concluded, based in part on further discoveries by Robert Broom in South Africa, that *Australopithecus* was not an ape, but a two-legged, upright-

A reconstructed model of an adult female Australopithecus africanus, *who probably lived about 2.5 million years ago. Her remains were found in South Africa.* (The Natural History Museum, London)

walking hominid. Leakey, however, had always maintained that the *Homo* species had been around a very long time and that Broom and Dart's australopithecines were not old enough to be ancestral to humans, a point on which few paleontologists agreed with him at the time.

But even Leakey became excited as they examined the site further. They had found what is referred to in paleoanthropology as a living floor, where thousands of years before a group of hominids had camped. It was older and less disturbed and contained more evidence than any living floor discovered up until that time.

Located about 300 miles west of the Indian Ocean, near Lake Victoria in Tanzania, the Olduvai Gorge cuts 300 feet deep or more, slashing across an ancient lake basin, now dry and dusty. The layers of deposits exposed along its walls cover a timespan of 1.9 million years—from the early Pleistocene epoch. And as it cuts through what used to be the shoreline of the ancient lake, the gorge exposes campsites, layer upon layer of them, of ancient hominids where they used to live along the edge of the lake. As the lake had risen and fallen in the past, hominids had camped by its shores, and then moved away, leaving their debris, which was covered by water and sediment and sometimes, by luck, a layer of volcanic ash from a nearby erupting volcano. Although the organic remains rotted away, the bones and stones

133

used as tools were protected from the erosion of sun and wind and were preserved.

At this site, bones of animals the hominids had consumed remained nearby, many of them broken to extract the marrow. And the area was strewn with stone tools—all this surrounding the skull that Mary had found. What's more, the volcanic layers of rock made it possible to use a new type of dating technique, potassium-argon dating, which was used for the first time on Mary's find. "Zinjanthropus," as Louis named it, was 1.75 million years old.

TOOLS OF SCIENCE: METHODS OF DATING

How do you tell how old a fossil is—since there's no date stamped on the back? In the past few decades, several different dating techniques, most of them highly sophisticated, have been developed. Many of them rely on the fact that radioactive isotopes have a predictable half-life, and the first person to come up with such a method was an American chemist named Willard Frank Libby (1908–80), who invented the method known as carbon-14 dating in 1947.

The carbon-14 isotope, discovered in 1940, has a surprisingly long half-life (the time it takes for half a radioactive substance to decay) of 5,700 years, and Libby saw how this fact could be put to use for dating. Cosmic rays convert some of the nitrogen-14 in the Earth's atmosphere into carbon-14, and the chemical process works in such a way that a certain amount of carbon-14 is therefore always present in the atmosphere. Since plants absorb carbon dioxide during photosynthesis, Libby figured, the molecules in all plant tissues would naturally contain a percentage of carbon-14 that occurs naturally in the atmosphere—a very small percentage, but it's always there. Since carbon-14 is radioactive, it's possible to measure the beta particles it emits with great precision.

Once a plant dies, though, photosynthesis stops, the plant no longer continues to absorb carbon dioxide, or carbon-14, and the carbon-14 that was present at the moment the plant died continues to break down without being replaced. Libby figured out that, by measuring the concentration of carbon-14 in a once-living plant, one could figure out how long the plant had been dead. The accuracy of the method was amazing and it could be used on samples of wood, parchment, cloth—anything that was made from a plant—to date objects up to 50,000 years old (70,000 years with special techniques). Because animals eat plants (or eat other animals that eat plants), the carbon-14 dating method can be extended to fossilized bones, as well. In the study of fossils, this is one of the few dating techniques that makes use of the fossil itself.

The presence of the tools meant, by definition, that this was an early human. This was the oldest hominid discovery yet made that was associated with tools, and the find radically changed scientific views about the timescale of human evolution. Most scientists had not thought australopithecines were capable of making tools—and Louis Leakey didn't want to accept their presence in the direct line of human history. Yet there it was.

In keeping with his views, Leakey created a new genus, *Zinjanthropus*, meaning "man from East Africa" (*Zinj* is the ancient name for East Africa,

---

We now know that carbon-14 concentrations in the atmosphere vary, depending upon the level of cosmic ray radiation, so that carbon-14 dating has to be calibrated against some known standard, such as annual growth rings in trees.

Because carbon-14 dating is limited in range, some method was needed for use when dating very ancient hominid fossils. Potassium-argon dating, discovered in 1961, is useful for dating rock that has a volcanic origin, and it can be used to test the rock surrounding a fossil found in volcanic areas such as the Great Rift Valley of East Africa. This advantage in dating in a volcanic area is one of the reasons the Great Rift Valley's fossils are so much easier to date than those found in the cave areas farther south.

When a volcano erupts, it spews forth molten lava, which includes minerals containing potassium. A tiny proportion of all potassium consists of its radioisotope, potassium-40, and the same kind of system can be used as with carbon-14; potassium-40 decays at a known rate to form the gas argon. In molten lava from a volcano, as the lava cools, crystals form, and the argon formed by the potassium-40 decay becomes trapped inside the crystals. By measuring the relative proportions of potassium-40 and argon in the crystals, it's possible to find out how long ago the lava cooled. There are also ways of checking to see if argon has been lost from the crystals.

In an area like the Great Rift Valley, layers composed of lava flows are typically interspersed with sediment, where the fossils are found, and a series of dates can be calculated at a series of intervals throughout the formation.

These and other dating techniques—such as thermoluminescence dating, fission-track dating, faunal dating (based on the stage of evolution of types of animal fossils found in the same layer), and measurements of remnant magnetic orientation—have added to the researcher's bag of tricks for dating very old fossils.

and *anthropus* means man.) The full name of this specimen became *Zinjanthropus boisei*, after a benefactor, Charles Boise, who had made it possible for the Leakeys to work the site. (Leakey was a tireless fundraiser and later took the Zinj skull with him to the United States and used it to persuade the National Geographic Society to back him.) "Nutcracker Man," as he was also sometimes called, had a distinctive ridge on the top of his head, where large muscles attached to move its powerful jaws and huge grinding molars (the reason for the name). Leakey eventually lost the nomenclature battle, however, and the specimen was later reclassified as *Australopithecus boisei*.

Meanwhile, as the Leakeys had been working at their sites in the Olduvai in Africa, just a few years before Mary's discovery of Zinj, a furor had taken place in the sedate halls of the British Museum in London: A scandal was afoot in the world of paleoanthropology.

PILTDOWN MAN REVISITED

For British paleontologists and anthropologists in 1911, the discovery of the Piltdown skull had been one of the most exhilarating, though baffling, of all the hominid fossils discovered up to that date. Found in a gravel pit near a place called Piltdown Common in Sussex, England by workmen building a road, the skull first came to the attention of a lawyer named Charles Dawson. Thinking that it surely must be a fossil, he took it to Arthur Smith Woodward, a paleontologist at the British Museum. Woodward was struck by the fact that, unlike the Neanderthal, a 30,000–40,000-year-old fossil discovered in Germany in 1857, this skull did not have prominent brow ridges and looked much more like modern humans. Intrigued, he went with Dawson to the site to see if they could find additional parts to the skeleton. After searching for several days, Dawson found a jawbone not far from where Woodward was also looking. The jawbone had the same brownish color as the rest of the skull, and seemed to be part of the same skeleton. But, while the skull looked strikingly like modern human skulls, the jawbone looked ape-like. Yet the teeth were not ape-like, but were ground down from chewing, like human teeth.

Tremendous controversy ensued, as investigators tried to explain some of Piltdown's strange characteristics. Based on the sketchy evidence of two fragmentary pieces—the face was missing and so were some of the hinge pieces for the jaws—Woodward put together a reconstructed skull that modeled the shape of the head and size of the brain capacity to fit with the ape-like jaw. Meanwhile, Sir Arthur Keith, a renowned anthropologist, preferred to emphasize the large brain capacity and apparently more humanlike structure of the cranium—a clearly more desirable heritage than

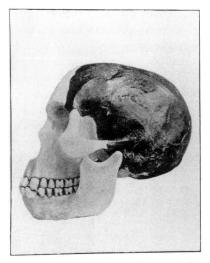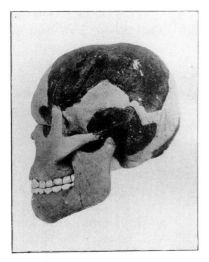

Two reconstructions of the Piltdown skull: Woodward's, with an ape-like jaw and small brain capacity (left), and Keith's, with both jaw and brain capacity similar to that of humans (The Natural History Museum, London)

Neanderthal—and fitted the jaw to look less ape-like. This, surely, was the link between apes and humans.

In fact, Keith liked the idea of a large-brained ancestor so much that, when Raymond Dart made his australopithecine find in Africa in 1924, Keith strongly questioned that it could be part of the line—having, as it did, a small brain and a humanlike jaw, just the opposite of Keith's reconstruction of Piltdown. And Keith was a well qualified opponent in these matters. He once invited doubters to smash a known skull, whereupon he reconstructed it perfectly from just a few fragments.

Not until the early 1950s, however, did the truth become certain about Piltdown, when Kenneth Oakley of the British Museum came up with a new method for measuring whether ancient bones were the same age (a method that is highly useful for any paleontological dig). When Oakley tested the jaw and skull of Piltdown with his radioactive fluorine test, he discovered that the skull was many thousands of years older than the jaw. Unless, as one wry scientist put it, "the man died but his jaw lingered on a few thousand years," the two fragments could not have belonged to the same individual.

Closer examination showed more. The jaw, it turned out, belonged to an orangutan and had been doctored, the teeth filed to look human and broken at the hinge to obscure its identity. Both fragments had been painted with brown stain and planted together in the gravel pit.

In November 1952 Kenneth Oakley announced his findings, and immediately the headlines blared: PILTDOWN APE-MAN A FAKE—FOSSIL HOAX MAKES MONKEYS OUT OF SCIENTISTS.

No one knows for sure who the culprit may have been who pulled off the hoax. Dawson may have been a good candidate: He had always been fascinated by fossils, and Piltdown did put his name in the science history books. Arthur Conan Doyle, the creator of Sherlock Holmes, is another possible suspect: He lived nearby and had a rascally liking for taunting scientists. But no one knows for sure. What is certain is that a prominent group of British scientists believed it was real—maybe because they were tired of seeing France and Germany get all the credit (at that time) as the cradle of early humanity, or maybe in some cases because the large brain echoed their own belief that the brain was the distinguishing trait that had spelled success for modern humanity.

It's easy from our vantage point to be amazed that such eminent and erudite scientists would fall for such a (now) obvious hoax. But hindsight, as Isaac Asimov once wrote on this subject, is cheap. Little was known in 1911 about the evolution of hominids—and none of the sophisticated dating and testing techniques now available were at the disposal of these investigators. Piltdown stands as an excellent lesson in how easy it is to be fooled by one's own biases and assumptions. We too often see what we want to see—and if we want to find truth we have to look much farther than that.

The immediate effect for paleoanthropology was disgrace. One member of Parliament suggested dismantling the British Museum. Sir Arthur Keith died in 1955, embarrassed, an otherwise fine career badly marred. The entire scientific validity of the profession was called into question as people wondered how so much could be surmised accurately from so little.

But science is designed to be self-correcting, to learn from mistakes. Before World War II, most anthropologists thought that a large brain was one of the prime distinguishing characteristics of hominids. Human vanity and the Piltdown hoax had gone a long way toward reinforcing this idea. Now that Piltdown man was discredited, more anthropologists paid attention to the mounting evidence that the small-brained australopithecines had walked erect. And the definition of the hominid now became focused less on the size of the brain than on the shape of the legs and pelvis. Not everyone has agreed on this premise, though, and cranial capacity still held sway, as well as the use of tools.

"HANDY HUMAN"

In 1961, only shortly after publication of the news of the "Nutcracker Man," which had fit so badly with Louis Leakey's view of human prehistory, his son Jonathan found a fragmentary skull at Olduvai Gorge that appeared to be an "advanced" hominid. Experts were able to place the date at 1.75 million years old by potassium-argon dating, and tools found in the area seemed to suggest the ability to use hands skillfully. So the fossil remains were named *Homo habilis*, or "Handy Man."

"This was the first evidence," the Leakeys' son Richard wrote, "that early members of the human lineage were contemporaries of the australopithecines, not descendants as was generally believed."

But debate rose up over whether this was really a distinct species or another australopithecine. Other fragments were found in the area that the Leakeys included as part of the same specimen, but some critics contested some of their identifications and assumptions.

In 1972, Bernard Ngeneo, who was working with Richard Leakey, found a similar skull, known as "1470," at East Turkana in Kenya. More pieces of this skull turned up, and Richard's wife, Meave, and a team of anatomists pieced it together. Richard Leakey believes that this skull, with its fairly large cranial capacity and dated at 1.89 million years old, is the oldest fossil of a true human ancestor (with australopithecines and other hominid fossils as side branches).

Meanwhile, more australopithecine evidence was building up elsewhere.

LUCY

On November 30, 1974 anthropologist Donald Johanson made one of the most exciting fossil discoveries ever made. He and his colleagues working at Hadar, in the Afar region of Ethiopia, had unearthed the most complete skeleton from 3.5 million years ago ever found. Judging from the shape of the pelvis, the field team made the assumption that the specimen is of a woman, and they named her Lucy. Although it is not absolutely certain she is a woman (there are no male pelvises around for comparison), the name stuck. The scientific name given "her" is *Australopithecus afarensis*, meaning "southern ape from Afar." (Even though none of these fossils is from the south, the first australopithecine was discovered in South Africa. Some people would also debate whether "ape" is the right description, but the name has stuck.)

Lucy was small of stature—standing only four feet tall—and had a small brain, reinforcing the concept that "walking tall" preceded enlarged brain capacity. The idea that a fossil woman had been found gave Johanson's discovery wide public appeal. As usual, however, there ensued many debates: about the age of the fossil (maybe it's only 3 million years old), over whether or not a fossil found by Mary Leakey 1000 miles away was the same species

(Johanson claimed it was, while she claimed it wasn't), and about the fossil's place in hominid history.

At the beginning of the 1980s, considerable confusion reigned and, while the finds of the 1980s were exciting, they did nothing, really, to clear it up.

TURKANA BOY

One day in August 1984, while working near Lake Turkana in Kenya, Kimoya Kimeu was retracing his steps to look at a rhinoceros skull he had passed up the previous day. Kimeu was the foreman of a group of experts working with Richard Leakey, who had become the director of the Kenya Museum, and Alan Walker of the Johns Hopkins Medical School. Sharp eyes and the ability to recognize odd and telling shapes, colors or textures are among the great skills required for this kind of work, along with the patience to retrace, relook and rethink everything. Something odd caught Kimeu's eye: a small skull fragment. The next day, the whole crew went back, and by dusk they had found it: additional scraps of a *Homo erectus* skull. They sieved and sorted and glued broken pieces together for a few more days. But large sections were still missing. They decided to give it one more day. The next day, by sheer luck, a stroke of a tiny dental pick disturbed soil and stones by the root of a thorn tree and the missing facial bones appeared. The team pressed on. By the time they were finished they had found the most complete skeleton of *Homo erectus* to date, and the first recovered of such great antiquity: 1.6 million years old. It was the skeleton of a boy, about 12 years of age, who stood about five feet four inches tall—surprisingly large for his age.

THE BLACK SKULL

Harking back to Mary Leakey's 1959 "Nutcracker Man," in 1985 Alan Walker discovered another australopithecine with huge jaws and pronounced ridges or crests. The "Black Skull"—so called because of the color of the fossil—was found near Lake Turkana in Kenya and has been dated between 2.5 and 2.6 million years old. Sometimes referred to as "hyperrobust," this older robust specimen was classified as *Australopithecus robustus*. It had the smallest brain capacity seen to date in a hominid skull, although it is not the earliest hominid ever found. Its bony ridges or crests are the most prominent ever found. Judging from the great grinding molars (four to five times as large as those of modern humans), the huge jaw muscles that attached to the crests apparently were used for chewing tough plant stalks or for cracking seeds or pits. The head and face were thrust farther forward than in other australopithecines, as well.

The fact that the Black Skull is three-quarters of a million years older than *A. boisei* suggests that robust australopithecines were a long-lived and successful branch of the hominid family, not a dead end. And, in fact, by 1989 Richard Leakey had put distance between himself and his original ideas about lineage, coming to the conclusion that, based on evidence, it's too early yet to tell much.

UNSOLVED MYSTERIES

The fossil record, though still sketchy, has become clearer on one point, though, in the last few decades: Australopithecines have to be recognized as a subfamily that shared important characteristics. They were habitually erect, bipedal beings with human-like teeth and relatively small brains. They used tools. And they were good candidates for the role of ancestors for humankind. The discoveries in East Africa and the recognition that the earliest humans and australopithecines occupied the same territory at the same time have raised many questions about the human family tree. Could australopithecines and humans have intermarried? Did humans eradicate the australopithecines? What happened? At the very least, we can be sure that the known australopithecines and humans must have a common ancestor (possibly, some anthropologists now think, *A. afarensis)*.

Many experts think we can reasonably postulate that *Homo* was present in East Africa by at least 2 million years ago, developing in parallel with at least one line (*A. boisei*) of australopithecines. These two hominid lines apparently lived at the same time for at least a million years, after which the australopithecine lineage apparently disappears forever. At the same time, the *Homo habilis* line was emerging into a later form, *Homo erectus*, which, in turn, developed into *Homo sapiens* (modern humans).

It seems increasingly likely that the family tree branched out about 6 million years ago, after which several types of apes, humans, human-apes and near-humans all coexisted for a long time. We still can't be sure from the evidence whether they lived together peaceably or as enemies—and we may never know the answer to this question. We also can't be sure, as the record now stands, which are our direct ancestors and which are just distant cousins.

There is a lot more of the story to come, and there is sure to be a lot more controversy along the way, as the search for more clues to the human and pre-human past continues.

E P I L O G U E

Science in the 20th century has come far and continues to move farther, into new and exciting territory. Doing science in the last five decades has, more than ever, become a voyage into strange realms made all the more strange and exciting because they are the very essence of the universe in which we live. And understanding these new realms—the world of sub-atomic particle physics, of DNA, of viruses, of the Big Bang and the inflationary universe—provides a window that allows us to see beyond ourselves and our own interests, to become more informed citizens of the universe and a more harmonious part of it. From this understanding we also come to value what those before us have discovered, to learn from it, and, always, to question, because this power—the power to use our minds—is the greatest gift that we can give, both to ourselves and to the world—the very essence of being human.

In a real sense the future of the world lies in the future of science. We depend upon knowledge to make informed decisions, and we need the fruits of science to gain that knowledge. We as citizens hold the reins. Because much (though not all) science now requires both coordinated teamwork and expensive equipment, science needs support from governments and industry in order to pursue the many big questions that still remain unanswered. The direction provided by us as citizens, taxpayers, stockholders and consumers will decide the future.

As the dawn of the 21st century approaches, many hard choices need to be made. Informed decisions about the use of scientific knowledge, more-over, require a sophisticated understanding of consequences and ethics. We stand at a juncture of time that is more exhilarating than any other in its possibilities. And whether or not we choose to do science ourselves we *all* have become participants in a great enterprise.

C H R O N O L O G Y

1945   May: World War II ends for Europe when Germany surrenders; Germany and Austria are divided by Allies into four zones

1945   August: Japan surrenders, ending World War II in the Pacific, after atomic bombs are dropped on Hiroshima and Nagasaki causing widespread death, destruction and injury

1945   The United Nations is created

1945   German rocket pioneer Wernher von Braun emigrates to the United States, where he continues his work, which lays the groundwork for establishment of the U.S. space program in the late 1950s

1946   The term *lepton* is introduced by Abraham Pais and C. Moller, referring to lightweight subatomic particles not affected by the strong nuclear force

1947   Hungarian-British physicist Dennis Gabor develops the basic concept of holography (though it is not truly practical until after the invention of the laser)

1947   Willard Frank Libby develops technique of radiocarbon dating

1947   The two-meson theory is rediscovered by R. Marshak and Hans Bethe, and, independently, by S. Sakata and T. Inoue; they realize that the meson in cosmic rays (now called a muon) is not the same as the one predicted by Yukawa (now called a pion)

1947   Cecil Frank Powell and his team in England isolate the pion (predicted in 1935 by Hideki Yukawa of Japan, who is credited with its discovery)

1947   The transistor is invented by a team of scientists at Bell Laboratories in the United States

1947   Under the leadership of nationalist Mahatma Gandhi, independence is won by India from British colonial rule

1948   Mahatma Gandhi is assassinated in January by a Hindu fanatic

1948   Richard Feynman, independently of Julian Seymour Schwinger and Shin'ichiro Tomonaga, develops renormalizable quantum electrodynamics (QED); Tomonaga is actually first, with his work completed in 1943, during World War II, but his work does not become known outside Japan until later

1948 George Gamow, Ralph Alpher and Robert Herman develop the Big Bang theory of the origin of the Universe

1948 Dedication of the 200-inch telescope on Palomar Mountain near Pasadena, California

1949 Soviet Union tests atomic bomb; by 1951, backyard bomb shelters are being sold in the United States

1949 North Atlantic Treaty Organization (NATO) is formed to control Soviet aggression

1949 The People's Republic of China is established after the triumph of Communist leader Mao Zedong in civil war in China

1950 Korean War begins

1951 Ten million U.S. homes have TVs

1951 UNIVAC, the first commercial computer, is accepted by the U.S. Bureau of the Census

1951 The field ion microscope is invented; it is the first instrument that can picture individual atoms

1951 First experimental breeder reactor built, in Idaho

1952 H-bomb (hydrogen fusion bomb), vastly more powerful than the atomic fission bomb, is tested by the United States in the Marshall Islands

1952 American physicist Charles H. Townes invents the maser

1952 U.S. Atomic Energy Commission builds the first breeder reactor, which produces plutonium at the same time it produces energy

1952 First accident at a nuclear reactor, Chalk River, Canada, where the nuclear core explodes due to a technician's error

1952 Contraception "pill" developed

1953 Francis Crick and James Watson discover the double-helix structure of the DNA molecule

1953 Murray Gell-Mann proposes the strangeness property of some subatomic particles

1953 Physicist Donald Glaser invents bubble chamber to detect subnuclear particles

1953 Piltdown man discovered to be a hoax, 41 years after the discovery of the fossils

1953 Korean War ends; North and South Korea remain divided

1954 Polio vaccine developed by Jonas Salk; with the help of another vaccine developed by Albert Sabin, reduces the number of polio cases in the United States by 96% by 1961

1954 CERN (originally, the Conseil Européen pour la Recherche Nucléaire, now known as the European Organization for

	Nuclear Research) is founded, located near Geneva, Switzerland
1954	Luis Walter Alvarez constructs a liquid hydrogen bubble chamber in which he can see tracks caused by subatomic particles
1954	Chen Ning (Frank) Yang and Robert Mills develop a gauge symmetrical field theory, a major step toward viewing the universe in terms of underlying symmetries that were broken in early cosmic evolution
1954	Walter Baade and Rudolph Minkowski identify the radio source Cygnus A with a distant galaxy
1954	In the case *Brown* vs. *Board of Education*, the U.S. Supreme Court rules that segregated schools (separate schools for white and black students) are unconstitutional
1954	Joseph McCarthy, U.S. senator from Wisconsin, conducts hearings many call a "witch hunt," trying to uncover the existence of communist infiltration in American government and life
1954	Nuclear submarine U.S.S. *Nautilus* is launched
1954	First frozen TV dinners introduced
1955	Albert Einstein dies
1955	Rosa Parks, an African-American woman in Montgomery, Alabama, refuses to give up her seat on a bus to a white man, as required by local law; she is arrested, and African Americans boycott Montgomery buses in protest for more than a year
1955	McDonalds and Disneyland open
1956	Yang and Tsung Dao Lee theorize that parity is not conserved in weak interactions—i.e., that the weak force does not function symmetrically. Experiments conducted by Chien Shiung Wu and collaborators the same year confirm their prediction
1956	Neutrinos are observed for the first time (originally predicted by Wolfgang Pauli in 1930)
1956	Hungarian uprising suppressed by Soviet troops
1957	October: Soviet Union launches *Sputnik*, the first artificial Earth satellite
1957	Schwinger proposes that the electromagnetic and weak forces are aspects of a single variety of interaction
1957	The Civil Rights Act is passed in the United States, and the Civil Rights Commission is formed to monitor civil rights issues; federal troops in Little Rock, Arkansas enforce school desegregation ordered by the Supreme Court
1958	Jan Hendrik Oort and colleagues use radio telescopes to map the spiral arms of the Milky Way galaxy
1958	*Explorer I*, the first U.S. satellite, is launched into orbit

1959	Antarctic Treaty is signed among 12 nations who agree to set aside political claims for 34 years and use the continent as a giant scientific lab
1959	Revolution in Cuba; Fidel Castro ousts the current dictator, seeks support from Soviet Union and forms communist government, causing a break in U.S.-Cuba relations
1959	Soviet Union sends probes to the Moon
1960	First quasars, the most luminous objects in the universe, are discovered by Allan Sandage and Thomas Matthews
1960	American physicist Theodore H. Maiman demonstrates the first successful laser
1960	Luis Walter Alvarez announces his discovery of resonances, very short-lived particles
1960	U.S. U-2 spy plane is shot down over Soviet Union
1960	Sixteen African nations gain independence
1961	Murray Gell-Mann and others develop what Gell-Mann calls the Eightfold Way to classify heavy subatomic particles
1961	Robert Hofstadter discovers that protons and neutrons have a structure
1961	Yuri Gagarin (Soviet Union) becomes first person to orbit Earth
1961	Berlin Wall is erected, dividing East and West Berlin, to prevent travel by East Germans into West Berlin
1962	A group headed by G. Danby at Brookhaven, New York discovers there are two types of neutrinos, one associated with the electron and a second associated with the muon (a third is now believed to exist, known as the tau neutrino)
1962	The term *hadron* is introduced by L. B. Okun to represent the group of particles affected by the strong nuclear force (such as protons, neutrons, pions and kaons)
1962	United States launches communications satellite Telstar
1962	Cuban missile crisis
1962	John Glenn becomes first U.S. astronaut to orbit Earth
1963	Maarten Schmidt finds red shift in the spectral lines of a quasar, indicating that quasars are the most distant class of objects in the universe
1963	Vaccine against measles perfected
1963	U.S. president John F. Kennedy is assassinated on November 22 in Dallas. Accused assassin Lee Harvey Oswald is shot and killed two days later by nightclub owner Jack Ruby
1964	Murray Gell-Mann and George Zweig introduce the concept of quarks

1964	Discovery of the omega-minus particle at Brookhaven National Laboratory confirms Gell-Mann's prediction of the "Eightfold Way"
1964	A fourth quark species, charm, is introduced by Sheldon Glashow and James D. Bjorken
1965	African-American leader Malcolm X is assassinated on February 21
1965	Stanford Linear Accelerator Center (SLAC) starts up in California
1967	Chia Lin and Frank Shu show that the spiral arms of galaxies may be created by density waves propagating across the galactic disk
1967	Jocelyn Bell and Anthony Hewish discover pulsars, leading to verification of the existence of extremely dense "neutron stars"
1968	Although he helped build the Soviet hydrogen bomb, Andrei Sakharov speaks out in favor of nuclear arms reduction in the Soviet Union and writes an essay called *Progress, Coexistence and Intellectual Freedom*, which is published in London, an act for which he is barred from secret work and persecuted most of the rest of his life
1968	Experiments at the Stanford Linear Accelerator Center support the theory that hadrons are made of quarks
1968	U.S. civil rights leader Martin Luther King, Jr. is assassinated
1968	U.S. presidential candidate Robert F. Kennedy is assassinated
1968	Liberal democratic-socialist government of Czechoslovakia is crushed by Soviet troops
1969	Fermilab is founded in Batavia, Illinois
1969	English biochemist Dorothy Hodgkin works out the structure for vitamin B12 using an electronic computer, for which she receives the 1964 Nobel prize for chemistry
1969	Jonathan Beckwith and coworkers isolate a single gene for the first time
1969	U.S. astronauts land on the Moon
1970	Howard M. Temin and David Baltimore discover in viruses an enzyme called reverse transcriptase, which causes RNA to be transcribed onto DNA, a key step in the development of genetic engineering
1970	Har Gobind Khorana of the University of Wisconsin leads a team that completely synthesizes a gene for the first time, assembling it directly from its component chemicals
1972	Murray Gell-Mann presents the beginnings of quantum chromodynamics (QCD), linking quarks and color forces and three flavors of quarks
1973	U.S. astronauts dock with space station *Skylab*, a laboratory for solar observation

147

1973	A calf is produced from a frozen embryo for the first time
1973	Stanley H. Cohen and Herbert W. Boyer establish genetic engineering when they show that DNA molecules can be cut with restriction enzymes, joined together with other enzymes and reproduced by introducing them into the chromosome of the bacterium *Escherichia coli*
1973	Watergate scandal, involving unethical campaign practices, surrounds Richard Nixon's presidency
1973	Vietnam War ends
1974	Paul Berg leads a meeting of 139 scientists, sponsored by the U.S. National Academy of Sciences and 18 other nations, in which they express concern about the future of genetic engineering
1974	Richard Nixon resigns as president; Vice President Gerald Ford is sworn in as president and shortly thereafter pardons Nixon
1975	In an unusual gesture of cooperation, U.S. and Soviet spacecraft dock in space in the Apollo-Soyuz project and their crews meet and exchange greetings
1976	U.S. spacecraft *Viking I* and *Viking II* make soft landings on the surface of Mars to test for the presence of life, gather other data and send back photos
1976	First development of a functional synthetic gene, complete with regulatory mechanisms, by Har Gobind Khorana and colleagues
1976	Susumu Tonegawa shows that antibodies are formed by combination by groups of genes that move close to each other on a chromosome
1976	Supersonic travel introduced when the Concorde, a British-French enterprise, makes its first flight at two times the speed of sound, 1,500 mph
1977	U.S. Department of Energy is created
1977	A baby mammoth is found in the Soviet Union; it has been frozen for 40,000 years
1977	*Star Wars* movie premiers
1978	Scientists succeed in finding the complete genetic structure of the virus SV40
1978	United States reestablishes relations with China
1979	Hostage crisis begins in Iran when 66 U.S. citizens are seized from the U.S. embassy and held hostage, most of them not released until January 20, 1981, after extensive attempts to obtain their release
1979	Nuclear reactor at Three-Mile Island in Pennsylvania comes close to having a core melt-down

1979	Revolution in Nicaragua deposes dictator Anastasio Somoza Debayle and establishes the Sandinista National Liberation Federation in power; eight years of civil war follow between SNLF under Daniel Ortega and U.S.-funded "contra" guerrilla forces
1979	Revolution in Iran establishes religious extremist Ayatollah Ruhollah Khomeini as leader
1979	Soviet troops are sent to Afghanistan to quell an uprising; many western nations boycott the Olympic Games in Moscow to protest the move
1980	First publication of a photo of an individual atom, taken at the University of Heidelberg in Germany
1980	Charles Weissmann of the University of Geneva reports the successful production of human interferon in bacteria
1980	Paul Berg, Walter Gilbert and Frederick Sanger share the Nobel prize for chemistry, Berg for his work on recombinant DNA and Gilbert and Sanger for methods to map the structure of DNA
1980	Mount St. Helens in Washington State erupts, providing new geological data about volcanoes and the interior workings of the Earth
1980	Iraq, under the leadership of Saddam Hussein, invades Iran, beginning an eight-year war, during which Iran sustains more than 700,000 casualties
1980	Lech Walesa in Poland establishes the first labor union, Solidarity, in a Soviet-bloc country
1980–81	First reported cases of AIDS (Acquired Immune Deficiency Syndrome)
1981	Alan Guth postulates that the early universe went through an "inflationary" period of exponential expansion
1981	U.S. hostages are freed in Iran
1981	Sandra Day O'Connor is appointed the first female judge on the U.S. Supreme Court
1982	Falklands War between Britain and Argentina
1982	First flight of the U.S. Space Shuttle
1982	First CD (compact disc) player, using laser technology to play music
1983	Electroweak unified theory verified in collider experiments at CERN; attempts accelerated to arrive at a unified theory of all four forces
1983	U.S. biologist Barbara McClintock receives the Nobel prize for physiology or medicine for her discovery of mobile genes in the chromosomes of corn

1983	Andrew W. Murray and Jack W. Szostak create the first artificial chromosome
1983	AIDS virus isolated
1983	Famine in Africa; years of drought result in widespread death from hunger, with one million dead in Ethiopia alone in the years 1984–85
1984	Prime minister of India, Indira Gandhi, is assassinated
1984	Alec Jeffreys discovers the technique of genetic fingerprinting, using core samples of DNA
1984	Steen A. Willadsen successfully clones sheep
1984	Allan Wilson and Russell Higuchi of the University of California at Berkeley become first to clone genes of an extinct species
1985	Mikhail Gorbachev takes office as general secretary of the Communist party in the Soviet Union; institutes a policy of "perestroika," involving sweeping economic changes, and "glasnost," a policy of openness toward other countries, including former cold-war opponents such as the United States, Britain, France and West Germany
1985	Hole in the ozone layer observed in satellite data from the area over Antarctica
1986	Entire crew of seven dies in *Challenger* space shuttle disaster
1986	U.S. Department of Agriculture issues the world's first license to market a living organism produced by genetic engineering, a virus used as a vaccine to prevent herpes in swine
1986	Soviet Union launches *Mir*, the first space station permanently occupied by human passengers
1986	Haitian dictator Jean-Claude Duvalier is overthrown and flees to France, ending 28 years of rule by the Duvalier family
1986	Ferdinand Marcos, who has imposed martial law in the Philippines since 1972, is forced to flee with his wife, Imelda, when he declares himself winner of a rigged election; Corazón Aquino takes office
1986	Cooling system fails at a nuclear plant at Chernobyl, 60 miles from Kiev in the Soviet Union; the core melts down and a radioactive cloud escapes. Immediate deaths mount to 30 and many more are afflicted with radiation-caused cancers. Soviet government attempts to seal the reactor by burying it under cement
1987	Proton-decay experiments in the United States and Japan detect neutrinos broadcast by a supernova in the Large Magellanic Cloud, ushering in the new science of observational neutrino astronomy

1988	Quasars are detected near the outposts of the observable universe; their red shifts indicate that their light has been traveling through space for some 17 billion years
1989	The Cosmic Background Explorer (COBE) is launched from the space shuttle to probe the cosmic background radiation
1989	Henry A. Ehrlich announces a method for identifying (or at least ruling out) an individual from the DNA in a single strand of hair
1989	Ya-Ming Hou and Paul Schimmel discover how to decode part of the genetic code found in transfer RNA
1989	Rally of tens of thousands of students for democracy in China is quelled in a bloody confrontation in Tiananmen Square in Beijing
1989	Bloody revolution in Romania
1989	Berlin Wall is torn down
1989	Demonstrations in Hungary, Czechoslovakia, Poland, Albania and Bulgaria lead to reform in those countries
1989	Loosened control by the communist regime in Yugoslavia leads to ethnic violence
1990	Hubble Space Telescope is launched
1990	Germany is reunited; first democratic elections held throughout Germany since 1932
1990	Iraq invades Kuwait
1991	The Gamma Ray Observatory (GRO) is launched from the space shuttle *Atlantis*
1991	The *Galileo* spacecraft takes a photo of Gaspra, the first photo ever taken of an asteroid in space
1991	Apartheid is dismantled in South Africa
1991	"Operation Desert Storm," a multinational military effort, pushes Iraq out of Kuwait
1991	Rajiv Gandhi, who succeeded his mother as prime minister of India, is assassinated
1991	Soviet Union breaks up after President Gorbachev resigns
1992	Members of the COBE science team announce that "ripples" have been detected in the cosmic background radiation
Dec. 1993	Two groups of researchers announce identification of gene for colon cancer
1994	Strong evidence for the existence of the "top" quark discovered at Fermi National Accelerator Laboratory

G L O S S A R Y

atom The smallest chemical unit of an element, consisting of a dense, positively charged nucleus surrounded by negatively charged electrons. The Greek thinker Leucippus and his student Democritus originally conceived of the idea of the atom in the 5th century B.C. as the smallest particle into which matter could be divided. (The word *atom* comes from the Greek word *atomos*, which means "indivisible.") But in the 1890s and early 20th century, scientists discovered that the atom is made up of even smaller particles, which are strongly bound together.

baryon One of a group of fundamental particles that theoretically are formed by three quarks joining together. (The term comes from the Greek word *barys* meaning "heavy.") Baryons include protons, neutrons, and several particles known as hyperons (lambda, sigma, xi and omega), which have very brief lifetimes and decay into nucleons and mesons or other lighter particles. For every baryon, there is an antibaryon.

chromosome A linear body found in the nucleus of all plant and animal cells that is responsible for determining and transmitting hereditary characteristics.

cosmic ray A stream of ionizing radiation that enters the Earth's atmosphere from outer space and particles (mostly muons) produced by collisions of these particles with the molecules of the upper atmosphere. The rays consist of high-energy atomic nuclei, muons, electromagnetic waves and other forms of radiation.

DNA (deoxyribonucleic acid) A nucleic acid that is the main constituent of the chromosomes of living cells; the molecule consists of two long chains of phosphate and sugar units twisted into a double helix with nitrogenous bases in between (adenine, thymine, guanine, and cytosine) joined by hydrogen bonds; the sequence of these bases determines individual heredity characteristics.

fundamental interaction One of the four forces found in nature: gravitation, electromagnetism, the strong nuclear force and the weak nuclear force. People have known about the first two for a long time, since they function over long distances and are easily observed, but the strong and weak forces were not discovered until the 20th century and their effects take place at the subatomic level.

fundamental particle An indivisible subatomic particle; a member of one of the groups of particles called quarks and leptons.

gene A hereditary unit that is located in a fixed place on a chromosome and has a specific influence on the physical makeup of a living organism. A gene is capable of mutation into various forms.

gluon The term used to describe the "glue," or tiny carrier particle, that mediates the strong force, binding together the subatomic particles called quarks that make up the atomic nucleus (protons, neutrons, mesons and so on). Gluons are theorized to exist in eight forms.

GUT (Grand Unification Theory) A unified field theory attempting to explain all the fundamental interactions (gravitation, electromagnetism and the strong and weak nuclear forces) under the umbrella of a single theory. Although no one has yet found a satisfactory theory, the search for it continues.

hadron One of a group of fundamental particles that take part in all four of the fundamental interactions. Hadrons are composed of quarks.

half-life The time it takes for one-half the nuclei in a given amount of a radioactive isotope to decay into another isotope or element. That is, after one half-life, one-half of the original sample of a radioactive element will have transformed. Half-lives of different isotopes range from a few millionths of a second to several billion years.

hominid A member of the family of two-legged primates that includes several human species, both extinct and living, among them *Australopithecus, Homo erectus, and Homo sapiens.*

Homo erectus The genus and species names given to Eugene Dubois's Java man, to Peking man and other extinct, ape-like humans. Large (six feet tall), with heavy browridges and a brain measuring one-half to two-thirds the size of the modern human brain, *Homo erectus* lived between .5 million to 1.6 million years ago.

lepton A class of fundamental particles that are governed by the weak interaction and are not affected by the strong interaction; there are 12 known leptons: the electron, electron neutrino, moun, muon neutrino, tao, tao neutrino and their antiparticles.

meson A subatomic fundamental particle of the hadron class (formed from a quark and its antiquark, according to current theory). One type of meson's existence was first suggested by Hideki Yukawa of Japan in 1935 to explain how the parts of the nucleus are bound together, and this meson, known as the pi meson or pion, was discovered in 1947; others have been discovered since.

muon A particle found in cosmic rays. The muon is classified as a lepton because it does not display the strong interaction; when a muon decays, an electron and a pair of neutrinos are left behind.

mutation Any alteration in an organism that can be inherited, carried by a gene or group of genes.

nucleon A proton or a neutron.

photon The packet that carries the electromagnetic force, having the properties of both a particle and a wave.

planetesimal Formational material out of which the planets are made.

QCD (See quantum chromodynamics)

QED (See quantum electrodynamics)

quanta Fundamental units, or "packets," of energy.

quantum chromodynamics (QCD) The theory that explains the strong nuclear force (carried by particles called *gluons*) in terms of its effect in binding the subatomic particles called quarks in different arrangements to form the particles found in the atomic nucleus. According to QCD, protons, neutrons and their antiparticles are composed of trios of quarks, while mesons are composed of one quark and one antiquark.

quantum electrodynamics (QED) In the 1920s and 1930s, Paul Dirac, Werner Heisenberg and Wolfgang Pauli proposed a set of equations that explained electromagnetic force in terms of the quantum nature of the photon. This set of equations, once it was developed further by Richard Feynman and others, provided the theoretical explanation for the interactions of electromagnetic radiation with atoms and their electrons and came to be called quantum electrodynamics or QED. QED succeeds in combining classical physics with quantum theory, explaining chemical reactions and much other observable behavior of matter and at the same time encompassing classic electromagnetic theory.

quark In nuclear physics, a subatomic particle, believed to be a fundamental nuclear particle, of which all hadrons are made. According to theory, quarks occur only in pairs or in threes, held together by the gluon, the carrier of the strong force. Quarks come in six types, known as "flavors," and the six flavors can also come in any of three "colors." Although quarks may have properties that make them invisible, enough evidence points to their existence that most scientists accept the theory.

RNA (ribonucleic acid) Found in all living cells, RNA consists of a single-stranded chain of alternating phosphate and ribose units with the bases adenine, quanine, cytosine, and uracil bonded to the ribose. The structure and sequence of bases in this chain are determinants of protein synthesis.

strong nuclear force   One of the four fundamental interactions found in nature, operating only at the subatomic level; the strong nuclear force binds the particles in the nucleus known as hadrons—the neutron, proton, unstable particles called hyperons and mesons (some of which exist only as "Virtual" particles).

subatomic   Describes units that make up the atom.

virus   A tiny agent of disease that consists of a core of nucleic acid surrounded by a protein coat. It is capable of replicating itself, but only in the presence of a living cell, and it produces disease by invading and destroying living cells and causing the release of a large number of new particles identical to the original one.

weak nuclear force   One of the four fundamental interactions found in nature, operating only at the subatomic level. The weak nuclear force is the force associated with radioactive decay and, together with the electromagnetic force, binds leptons (including electrons and muons).

F U R T H E R
R E A D I N G

BOOKS ABOUT SCIENCE IN GENERAL:

Cole, K. C. *Sympathetic Vibrations: Reflections on Physics as a Way of Life*. New York: William Morrow and Co., 1985. Well-written, lively, and completely intriguing look at physics presented in a thoughtful and insightful way by a writer who cares for her subject. The emphasis here is primarily modern physics and concentrates more on the ideas than the history.

Gardner, Martin. *Fads and Fallacies in the Name of Science*. New York: New American Library, 1986 (reprint of 1952 edition). A classic look at pseudoscience by the master debunker. Includes sections on pseudo-scientific beliefs in the 19th century.

Gonick, Larry, and Art Huffman. *The Cartoon Guide to Physics*. New York: Harper Perennial, 1991. Fun, but the whiz-bang approach sometimes zips by important points a little too fast.

Hann, Judith. *How Science Works*. Pleasantville, N.Y.: The Reader's Digest Association, Inc., 1991. Lively well-illustrated look at physics for young readers. Good, brief explanations of basic laws and short historical overviews accompany many easy experiments readers can perform.

Hazen, Robert K., and James Trefil. *Science Matters: Achieving Scientific Literacy*. New York: Doubleday, 1991. A clear and readable overview of basic principles of science and how they apply to science in today's world.

Holzinger, Philip R. *The House of Science*. New York: John Wiley and Sons, 1990. Lively question-and-answer discussion of science for young adults. Includes activities and experiments.

BOOKS ABOUT THE PHYSICAL SCIENCES, 1946 TO THE 1990s:

Ferris, Timothy. *The World Treasury of Physics, Astronomy and Mathematics*. Boston: Little, Brown, 1991. Historical writings, essays, poetry. A wonderful anthology for both the science student and the general reader, sparkling with appreciation and wonder. A good book to give not only to science students and enthusiasts, but also to those skeptics who claim that science takes the poetry out of life.

On Physics, Particle Physics and the Quantum:

Boorse, Henry A., Lloyd Motz, and Jefferson Hane Weaver. *The Atomic Scientists: A Biographical History.* New York: John Wiley and Sons, Inc., 1989. Interesting approach to the history of atomic science. The development is followed through a series of brief but lucid biographies of the major figures. May be a little difficult without some background.

Crease, Robert P., and Charles C. Mann. *The Second Creation: Makers of Revolution in 20th-Century Physics.* New York: Macmillan Publishing Co., 1986. Thorough, well done study, but sometimes a little heavy going for the average reader.

Einstein, Albert. *Out of My Later Years.* New York: Philosophical Library, 1950. Essays on science, humanity, politics and life by the gentle-thinking Einstein.

Farnes, Patricia, and G. Kass-Simon, eds. *Women of Science: Righting the Record.* Bloomington: Indiana University Press, 1990. Short one- to three-page profiles of women who—often little recognized—have made contributions in archaeology, geology, astronomy, mathematics, engineering, physics, biology, medical science, chemistry and crystallography. Written by scientists, this book offers a wealth of information and an eye-opener to the fact that many more women have made contributions in science than generally thought.

Feinberg, Gerald. *What Is the World Made Of?: Atoms, Leptons, Quarks and Other Tantalizing Particles.* New York: Doubleday, 1977. A good medium-level book on particle physics. Concerned primarily with physical processes rather than history or historical perspective. A little dated in some sections.

Feynman, Richard P. *The Feynman Lectures on Physics,* 3 volumes. Reading, Mass.: Addison-Wesley, 1963. Difficult for all but upper-level students, but this is a splendid series and Feynman at his finest.

———. *QED: The Strange Theory of Light and Matter.* Princeton: Princeton University Press, 1983. Vividly lucid and accessible to just about all interested readers. A slim and elegant example of a first-rate and original mind tackling a fascinating and difficult subject.

Glashow, Sheldon L. *Interactions: A Journey Through the Mind of a Particle Physicist and the Matter of This World.* New York: Warner Books, 1988. Part autobiography, part science history, and all Glashow with his strong ego intact. Charming enough though to get the reader over the difficult spots, both personal and scientific, and a good look at the life, and particularly the youth, of one of the 20th century's major physicists.

Gleick, James. *Genius: The Life and Science of Richard Feynman*. New York: Pantheon Books, 1992. An uncommonly perceptive look at one of the most original minds of modern physics, written with the insights of a master science writer.

Han, M. Y. *The Secret Life of the Quanta*. Blue Ridge Summit, Pa.: Tab Books, 1990. Coming at the quanta from the practical rather than the theoretical end, Han concentrates on explaining how our knowledge of the quanta allows us to build such high-tech devices as computers, lasers, and the CAT Scan—while not neglecting theory and the bizarre aspects of his subject.

————. *The Probable Universe*. Blue Ridge Summit, Pa.: Tab Books, 1993. Han follows up his successful *Secret Life of the Quanta* with a closer look at the mysteries and controversies, as well as the theoretical and technological successes of quantum physics.

Hazen, Robert K., and James Trefil. *Science Matters: Achieving Scientific Literacy*. New York: Doubleday, 1991. Includes good brief discussions on relativity and quantum physics.

Hazen, Robert. *The Breakthrough: The Race for the Superconductor*. New York: Summit Books Simon and Schuster, 1988. A straightforward, no-frills account. Nicely done, if not always exciting.

Herbert, Nick. *Quantum Reality: Beyond the New Physics*. Garden City, N.Y.: Anchor Press, 1985. Herbert plays with the paradoxes of quantum theory in this book and turns out a completely captivating and thought-provoking look at some of the more unsettling aspects of quantum physics. Very readable.

Hoffmann, Banish. *The Strange Story of the Quantum*. New York: Dover Publications, 1959. A now classic look at the birth and implications of quantum physics by a classy and commanding writer. If not as jazzy as some more recent attempts to cover the same material, it is a sane and sensible approach to a not always sensible subject. A warning, though: It was first published in 1959, so some of the material is now dated.

Holton, Gerald. *Thematic Origins of Scientific Thought*, revised edition. Cambridge, Mass.: Harvard University Press, 1988. A good, standard study but may be tough reading for the average reader.

Jones, Roger S. *Physics for the Rest of Us*. Chicago: Contemporary Books, 1992. Informative, non-mathematical discussions of relativity and quantum theory told as a straightforward matter, along with discussions on how both affect our lives and philosophies.

Lamer, Max. *Conceptual Development of Quantum Mechanics*. New York: McGraw-Hill, 1966. Superior study but tough going without some familiarity with the subject.

Lederman, Leon M., and David N. Schramm. *From Quarks to the Cosmos: Tools of Discovery.* New York: Scientific American Library, 1989. Takes a good look not only at the theorists but at the experimenters and experiments in the development of our understanding of the atom. Well-done and easy-to-follow book.

Lederman, Leon, with Dick Teresi. *The God Particle: If the Universe Is the Answer, What Is the Question?* New York: Houghton Mifflin, 1993. A sprightly and often humorously personal look at the history of physics and particle physics. Easy to read, lucid and informative.

McCormmach, Russell. *Night Thoughts of a Classical Physicist.* Cambridge, Mass.: Harvard University Press, 1982. A fascinating novel treating the mind and emotions of a turn-of-the-century physicist in Germany who attempts to understand the disturbing changes happening in physics and the world around him.

Miller, Arthur I. *Imagery in Scientific Thought.* Cambridge, Mass.: The MIT Press, 1986. Historical and philosophical look at the development in the thinking of physics. Might be a little abstract for the average reader.

Morris, Richard. *The Nature of Reality.* New York: McGraw-Hill, 1987. A tough-minded, critical and easy-to-read look at the direction the science of physics is going today.

———. *The Edges of Science: Crossing the Boundary from Physics to Metaphysics.* New York: Prentice-Hall, 1990. A tough-minded and critical look at the current problems of physics.

———. *The Nature of Reality.* New York: McGraw-Hill Book Company, 1987. Tough-minded, keenly written look at the various philosophical and scientific interpretations of modern physics and cosmology.

Pagels, Heinz R. *The Cosmic Code: Quantum Physics as the Language of Nature.* New York: Simon and Schuster, 1982. Quickly becoming a standard reference for the history and problems of quantum physics. Brilliantly written and easy to read, offering a clear look at some of the quantum paradoxes.

Pais, Abraham. *Inward Bound: Of Matter and Forces in the Physical World.* Oxford: Clarendon Press, 1986. Very well-done study but not always an easy read.

Peat, David F. *Einstein's Moon: Bell's Theorem and the Curious Quest for Quantum Reality.* Chicago: Contemporary Books, Inc., 1990. A popularly written study of some of the major paradoxes of quantum theory.

Riordan, Michael. *The Hunting of the Quark: A True Story of Modern Physics.* New York: Simon and Schuster, 1987. A lively and engrossing history aimed at the general reader.

Segrè, Emilio. *From X-Rays to Quarks: Modern Physicists and Their Discoveries.* San Francisco: W. H. Freeman and Co., 1980. Good study, sometimes tough going.

Shapiro, Gilbert. *A Skeleton in the Darkroom: Stories of Serendipity in Science.* San Francisco: Harper & Row, 1986. Engagingly told stories of scientific discovery, including the discovery of cosmic microwave background radiation, the discovery of the J/Psi meson, the development of the asteroid/comet impact hypothesis, and the discovery of pulsars.

Von Baeyer, Hans Christian. *Taming the Atom.* New York: Random House, 1992. Well-told, easy-to-read and up-to-date look at the history and development of atomic physics. Von Baeyer also takes a look at the most recent developments in the field and at present and upcoming attempts at experiments to help clear up the quantum mysteries.

Weinberg, Steven. *The Discovery of Subatomic Particles.* New York: Scientific American Library (An imprint of W. H. Freeman and Co.), 1983.

Weisskopf, Victor F. *The Joy of Insight: Passions of a Physicist.* New York: Basic Books, 1991. Delightful and personal reminiscences by an easygoing and thoughtful physicist who knew many of the great quantum pioneers first-hand.

———. *The Privilege of Being a Physicist.* New York: W. H. Freeman and Co., 1989. Reminiscences and random thoughts by a well-respected physicist.

On Astronomy and Cosmology:

Bartusiak, Marcia. *Through a Universe Darkly.* New York: HarperCollins, 1993. A sparkling account of the history of the universe as we now think of it and the story of the people who have unfolded it.

Boslough, John. *Masters of Time: Cosmology at the End of Innocence.* Reading, Mass.: Addison-Wesley, 1992. A very good critical look at the current problems in cosmology and a hard questioning look at the current theoretical search for the big answers.

Ferris, Timothy. *Coming of Age in the Milky Way.* New York: William Morrow, 1988. An exciting, readable account of some of the great discoveries of science from the time of the Greeks to the present. Draws on a deep appreciation for the beauty of science and its contributions to modern culture. Includes discussions of Big Bang theory, SETI, Guth's theories and other areas of modern cosmology.

Hawking, Stephen. *A Brief History of Time.* New York: Bantam Books, 1988. This lucid, personal and engrossing look at the "big picture" of modern physics by the brilliant scientist Stephen Hawking became a best-seller in 1988.

Lightman, Alan, and Roberta Brawler, eds. *Origins: The Lives and Worlds of Modern Cosmologists.* Cambridge, Mass.: Harvard University Press, 1990.

Overbye, Dennis. *Lonely Hearts of the Cosmos.* New York: HarperCollins, 1991. A highly readable look at the men and women in the forefront of modern cosmology.

BOOKS ABOUT THE LIFE SCIENCES, 1946 TO THE 1990s:

Asimov, Isaac. *A Short History of Biology.* Garden City, N.Y.: The Natural History Press, 1964. A brief overview of discoveries in biology from earliest times to the 1960s, written clearly, in a blow-by-blow style.

Chase, Allan. *Magic Shots.* New York: William Morrow and Company, Inc., 1982. The story of the quest for vaccinations to prevent infectious diseases in the 19th and 20th centuries, from Edward Jenner's smallpox vaccination to polio vaccines and the ongoing search for cancer vaccines.

Facklam, Margery, and Howard Facklam. *Healing Drugs: The History of Pharmacology.* New York: Facts On File, 1992. Part of the Facts On File Science Sourcebook series, this book tells the stories of Paul Ehrlich and his magic bullet, Best and Banting and their research on insulin, and Fleming's discovery of penicillin, as well as more recent drug research.

Kornberg, Arthur. *For the Love of Enzymes: The Odyssey of a Biochemist.* Cambridge, Mass.: Harvard University Press, 1989. One of the premier biologists of our time provides the 19th- and early 20th-century background for recent advances in biology, as background to his autobiographical account of his own significant work.

Moore, John A. *Science As a Way of Knowing: The Foundations of Modern Biology.* Cambridge, Mass.: Harvard University Press, 1993. A cultural history of biology as well as a delightful introduction to the procedures and values of science, written by a leading evolutionary-developmental biologist. Nontechnical and readable, it is excellent reading for the nonspecialist who wants to know more about how modern molecular biology, ecology, and biotechnology came to be.

Shorter, Edward, Ph.D. *The Health Century: A Companion to the PBS Television Series.* New York: Doubleday, 1987. An easygoing account of the scientific pursuit of keys to better health in the 20th century, set in the social context of the times.

On Genetics and DNA:

Bishop, Jerry E., and Michael Waldholz. *Genome.* New York: Touchstone Simon and Schuster, 1990. The story of the heroic attempt to provide a complete map of all the genes in the human body. Told as an adventure story and featuring a wide cast of interesting participants. Very readable and exciting.

Crick, Francis. *What Mad Pursuit: A Personal View of Scientific Discovery.* New York: Basic Books, Inc., 1988. Crick reminisces about the discovery of the double helix structure of DNA and in the process weaves a story of science, experiment and theory—in the discipline of biology, and molecular biology, in particular. Fascinating reading.

Keller, Evelyn Fox. *A Feeling for the Organism: The Life and Work of Barbara McClintock.* New York: W. H. Freeman, 1983. An unusually insightful biography of a highly independent scientist and her work, as well as a revealing look at the way science is done and the role played by those who buck the mainstream. Keller, who is trained in theoretical physics and molecular biology, brings a keen understanding of human dynamics in science to her account of this extraordinary renegade geneticist.

McCarty, Maclyn. *The Transforming Principle: Discovering That Genes Are Made of DNA.* New York: W. W. Norton and Company, 1985. A first-person account by one of the three scientists that discovered in 1944 that DNA is *the* genetic material, giving birth to molecular biology and modern genetics.

Moore, Ruth. *The Coil of Life: The Story of the Great Discoveries in the Life Sciences.* New York: Alfred A. Knopf, 1961. Engaging account of the events leading up to and beyond the discovery of the structure of DNA in the 1950s.

Newton, David E. *James Watson and Francis Crick: Discovery of the Double Helix and Beyond.* (Makers of Modern Science Series.) New York: Facts On File, 1992. Written for young adults, this dual biography traces the paths of Watson and Crick through the exciting years of their teamwork on the structure of DNA and also explores their later work, done separately.

Portugal, Franklin H., and Jack S. Cohen. *A Century of DNA.* Cambridge, Mass.: The MIT Press, 1977. Provides a thorough look at the background leading up to the discovery of the role played by DNA in heredity.

Watson, James D., and John Tooze. *The DNA Story: A Documentary History of Gene Cloning.* San Francisco: W. H. Freeman and Co., 1981. An off-beat and very compelling look at the history of DNA research told through the use of newspaper clips, personal correspondence, official documents and narrative. Well done and interesting, but it does require a little background on the part of the reader to help fit all the pieces in perspective.

Watson, James D. *The Double Helix.* New York: Atheneum, 1968. A now-classic, highly personal account of the search for the elusive structure of DNA, as engrossing now as when first written. Science research as a battlefront in one of the first and best "warts-and-all" looks at how science is done.

On Evolution:

Eldridge, Niles. *Life Pulse: Episodes from the Story of the Fossil Record.* New York: Facts On File, 1987. A little tough going and dry but well worth the effort for the serious student.

Milner, Richard. *The Encyclopedia of Evolution.* New York: Facts On File, 1990. Just what the title says, written in an easygoing, informal style that makes it a joy to use for research or just endless hours of browsing fun. Plenty of surprises, too, including looks at popular films dealing with the subject and lots of little-known oddities.

Reader, John. *The Rise of Life.* New York: Alfred A. Knopf, 1986. Beautifully illustrated popular-level book covering the origins and evolution not only of humans but of all life forms. Very well done.

On the Search for Human Origins:

Burenhult, Gören, gen. ed. *The First Humans: Human Origins and History to 10,000 BC.* Volume 1 of the Landmark Series from the American Museum of Natural History. New York: HarperSanFrancisco (A Division of HarperCollins Publishers), 1993. Originally published by Weldon Owen Pty Limited, McMahons Point, Sydney, Australia and Bra Böcker AB, Sweden, 1993. Beautifully organized, edited and illustrated. Charts, maps and illustrations all work hand in hand with a highly readable narrative to make this an indispensable up-to-date volume for any personal or school library.

Lambert, David, and the Diagram Group. *The Field Guide to Early Man.* New York: Facts On File, 1987. Well-designed and carefully illustrated "handbook" on all aspects of early humans. A wonderful book for finding facts, or just reading for fun. Easy to read and endlessly fascinating.

Lewin, Roger. *Bones of Contention: Controversies in the Search for Human Origins.* New York: Simon and Schuster, 1987. Although primarily concerned with the controversies and developments in the latter half of the 20th century, this book does have good, readable background material about the search for human origins during the period from 1895 to 1945, as well.

Spencer, Frank. *Piltdown: A Scientific Forgery.* London: Natural History Museum Publications, Oxford University Press, 1990. Exhaustive but highly readable look at the Piltdown hoax and its discovery. Covers just about everyone involved in the story, including the most likely suspects. Illustrations and seldom-seen photographs help bring the story to life. A must for anyone interested in science and the Piltdown mystery.

Willis, Delta. *The Hominid Gang: Behind the Scenes in the Search for Human Origins.* Introduction by Stephen Jay Gould. New York: Viking, 1989. Fascinating, on-the-spot examination of how anthropology is done in Africa by Richard Leakey and his crew of trained specialists, recorded with a skeptical eye by photojournalist Delta Willis.

MAGAZINES:

Discover: The World of Science
500 South Buena Vista Street
Burbank, CA 91521
  Subscriptions:
Discover
P.O. Box 420105
Palm Coast, FL 32142–0105
800–829–9132

An excellent popularly written magazine covering all aspects of current science, from black holes to anthropology—with lively full-color illustrations.

Natural History
American Museum of Natural History
Central Park West
New York, NY 10024
  Subscriptions:
Natural History
P.O. Box 5000
Harlan, IA 51537–5000

Published by the American Museum of Natural History, the best popular coverage of the life sciences around—beautifully illustrated with full color.

Science News
The weekly newsmagazine of science
231 W. Center St.
P.O. Box 1925
Marion, OH 43305
Weekly science news written in upbeat style.

Smithsonian
Smithsonian Associates
900 Jefferson Drive
Washington, DC 20560
    Subscriptions:
Smithsonian
P.O. Box 55593
Boulder, CO 80322–5593
800–766–2149
Fine source for panoramic and colorful views of the history of science, popularly written, with lush full-color illustrations.

M U S E U M S
A N D   O T H E R
S O U R C E S

American Museum of Natural History
Central Park West at 79th Street
New York, NY 10024
General Information: (212) 769–5100
Stunning exhibits of animals shown in their natural habitats, including the
    newest permanent exhibit, the Hall of Human Biology and Evolution.
    Also, Naturemax Theater shows films and the Hayden Planetarium
    offers sky shows and laser light shows, as well as a museum of astro-
    nomical photographs, exhibits and artwork.

The Astronomical Society of the Pacific (ASP)
390 Ashton Avenue
San Francisco, CA 94112
The ASP's purpose is to serve as a bridge between astronomers and the
    general public. A nonprofit group, the ASP publishes *Mercury*, a
    popular magazine on astronomy, and is an excellent source of informa-
    tion and educational materials on all aspects of the subject.

The Carnegie Museum of Natural History
4400 Forbes Avenue
Pittsburgh, PA 15213
General Information: (412) 633–3131
In the Discovery Room, visitors are welcome to touch and study at close
    range items such as starfish and bear's fur.

Exploratorium
3601 Lyon Street
San Francisco, CA 94123
(Inside the Palace of Fine Arts)
General Information: (415) 561–0360
Especially designed for curious kids and adults, this museum is known for
    its interactive science demonstrations. Other attractions include the

McBean Theater, which shows films on Saturday and Sunday, and the Tactile Dome, a large geodesic dome with 13 pitch-black rooms; visitors find their way out by the sense of touch only.

Field Museum of Natural History
Lake Shore Drive at Roosevelt Road
Chicago, IL 60605
General Information: (312) 922–9410
Includes the world's largest mounted dinosaur skeleton, a reconstructed specimen of the Brachiosaurus, 75 feet long and 45 feet high; Messages from the Wilderness exhibit recreates 18 wilderness parks and wildlife reserves in North and South America, showing animals in realistic settings, and includes videos, interactive computers and sounds; Inside Ancient Egypt exhibit includes a life-size tomb of a pharaoh from 2400 B.C. and 23 mummies.

Harvard University Museums of Cultural and Natural History
26 Oxford Street
Cambridge, MA 02138
General Information: (617) 495–1910
Includes four museums: The Botanical Museum, the Museum of Comparative Zoology, the Mineralogical and Geological Museums, and the Peabody Museum of Archaeology and Ethnology, located in the same building.

Liberty Science Center
251 Phillip Street
Jersey City, NJ 07305
(In Liberty State Park)
General information: (201) 200–1000
Houses special exhibits as well as three theaters: the Kodak Omni Theater, with movies projected on a huge overhead dome; the Joseph D. Williams Demonstration Theater, which has 3-D films alternating with live science demonstrations of communication, electricity and air pressure; and the Interactive Theater, where the audience interacts and affects the outcome of the films.

National Air and Space Museum
Sixth Street and Independence Avenue, S.W.
Washington, D.C. 20560
General Information: (202) 357–1686
In addition to exhibits on the history of flight and space exploration, the Samuel P. Langley Theater presents half-hour films, and the Albert Einstein Planetarium runs 30- to 40-minute shows continuously.

National Museum of Natural History
Smithsonian Institution
Tenth Street and Constitution Avenue, N.W.
Washington, D.C. 20560
General Information: (202) 357–2700
The O. Orkin Insect Zoo at the National Museum has the best live insect
    museum in the United States, while free lectures, films and perfor-
    mances are available at the Baird Auditorium.

The Natural History Museum
Cromwell Road
London SW7 58D, England
General Information: 44–71–938–9123
Attractions include the Permanent Dinosaur Exhibition, one of the world's
    best collections, with robotic models and hands-on exhibits; and the
    Blue Whale, with a full-size model of this giant mammal, the largest
    ever to live on Earth.

The Planetary Society
65 N. Catalina Avenue
Pasadena, CA 91106
Cofounded by astronomer Carl Sagan and planetary scientist Bruce Murray,
    this nonprofit group has 100,000 members and is devoted to explora-
    tion of the Solar System and the search for extraterrestrial life. For
    more information about the Planetary Society's SETI program, call
    1–800–9WORLDS, toll free.

Society for Amateur Scientists
4951 D Clairmont Square, Suite 179
San Diego, CA 92117
A group formed to encourage the idea that everyone can do science and that
    it's still possible to make valid contributions in science without an
    advanced degree and without a government grant or backing by a large
    institution.

University Museum of Archaeology and Anthropology
University of Pennsylvania
Spruce and 33rd Streets
Philadelphia, PA 19104
General Information: (215) 898–4000
Includes two special archaeological exhibits: Egyptian Mummy: Secrets in
    Science and Ancient Mesopotamia: The Royal Tombs of Ur.

I N D E X

Boldface numbers indicate major topics.
Italic numbers indicate illustrations.
The letter t indicates tables.

169

INDEX

Callisto (moon of Jupiter) 77
Caloris Basin 70
cancer 121–123 *See also* colon cancer; Kaposi's sarcoma; skin cancer
Cannon, Annie Jump 47
Cape Canaveral xii
carbon 43–44, 80, 118
carbon-14 dating (radiocarbon dating) 134–135, 143
carbon dioxide 66, 69, 73, 96–97
Carr, Michael 73
catastrophism 91, 93
CD4 T cells 124
CD (compact disc) players 149
Centers for Disease Control, U.S. (CDC) 123
ceramics 18
Ceres (asteroid) 74
CERN (Conseil Européen pour la Recherche Nucléaire) 21, 31, 36, 144–145, 149
CFCs *See* chlorofluorocarbons (CFCs)
Chadwick, James 5–6
Challenger disaster 13–14, 150
Chamberlain, Owen 18
Chang, Sherwood 114, 117
Chargaff, Erwin 105
charge, electric 29–32
charge center (of subatomic particle groups) 24
charm (property of quarks) 7, 31–32, 147
Charon (moon of Pluto) 82–83
Chernobyl disaster xv, 150
chickens 122
chlorofluorocarbons (CFCs) 96
Christy, James W. 83
chromosomes 152
  artificial, creation of 150
  discovery of 100
  gene shifts on 126–127
  genetic markers used to find 129
  in retroviruses 122
  structure of 101
  in viroids 120
civil rights movement xiv
clay, and the origin of life 117–118
cloning xv, 118–120, *119*, 122, 150
cloud chambers 15, 20
cobalt 60 28
Cocconi, Giuseppe 52
Cockcroft, John 20
Cohen, Stanley H. 130, 148
cold fusion 48–49
Cold War xi, xiii
colon cancer 151
color (property of quarks) 7, **32–33**, 147
color-force 33
comets 94
compact disc (CD) players *See* CD (compact disc) players
computers **8–9**, 13, 53, 144
Concorde (supersonic airliner) 148
continental drift **88–91**
contraceptives, oral 144
Cooper, Leon N. 18
corn, genetic transposition in 126–127
Cosmic Background Explorer (COBE) (NASA spacecraft) 56, 151
cosmic background radiation **55–57**
cosmic rays 14–15, 134–135, 143, 152
cosmological red shifts 49, 51
cosmology 37, 39
Cosmotron 20
Crab Nebula 40, 45

Cretaceous age 94, 97
Crick, Francis xvi, **102–110**, *103*, 113, 115, 127, 144
critical temperature 18
cryogenics 28
crystallography *See* X-ray crystallography
Cuban missile crisis xi–xii
Curie, Marie xv, 23, 110
Curie, Pierre xv, 23
Cuvier, Georges 91
cyclotron 20 *See also* synchro-cyclotron
Cygnus A 145
Cygnus XI 58
cystic fibrosis 129
cytology 100
cytoplasm 122–123
cytosine 105, *109*

D

Dalton, John xv, 4–5, 30
Danby, G. 146
dark matter **59–61**
Dart, Raymond 132–133, 137
Darwin, Charles x, 91–92, 132
data processing 9
dating techniques (for fossils) **134–135**
Davies, Kay 129
Dawson, Charles 136, 138
Delbrück, Max 101, 121, 128
Democritus xv, 2, 4–5
density waves 147
deoxyribonucleic acid (DNA) xvi, 152
  as architect of life 100
  double-helix structure of **101–110**, *107–109*, 144
  genetic markers based on 129
  identification methods based on 150, 151
  Nobel prizes awarded for research on (1980) 149
  restriction enzymes used to cut 128, 148
  retroviral production of (reverse transcription) 122, 147
deoxyribose 101, *109*
desktop computers 9
deuterium (heavy hydrogen) 114
dideoxynosine (ddI) 125
digital computing 8
dinosaurs 97
  extinction of 74, **94–95**
Dione (moon of Saturn) 77
Dirac, Paul 5
Doppler shift 39
Double Helix, The (James Watson) 108
double-helix model (of DNA) 101–110, 144
down quarks 31–32
Doyle, Arthur Conan 138
Drake, Frank 53
Duchenne muscular dystrophy 129–130
DuPouy, Gaston 121
dust 83–84

E

Earth (planet) **86–97**
earthquakes 89–90, *90*
Earthrise *86*
E. coli See Escherichia coli (*E. coli*)
Eddington, Arthur Stanley vii, 43, 47
Edison, Thomas Alva 8
Egyptians, ancient 38
Ehrlich, Henry A. 151
Ehrlich, Paul 95, 97
Eightfold Way 29–30, 146–147

INDEX

Heisenberg, Werner 5–6, 9, 11
heliosphere *See* solar wind (heliosphere)
helium
    as component of gas giants 75, 84
    as coolant 18, 40
    lithium transformed into 20
    in Neptune atmosphere 81
    produced within stars 43–44
    in Solar System formation 83
    studied by Feynman 13
hemophilia 124
hepatitis B vaccine 121, 130
heredity 100–101
Herman, Robert 144
herpes 150
Herschel, William 79
Hershey, Alfred D. 121, 128
Hess, Harry 89–90
heterosexual transmission of AIDS 124
Hewish, Anthony 45–46, 51, 147
Higuchi, Russell 150
Hitler, Adolf 43
HIV (human immunodeficiency virus) 120, 123–125
Hoagland, Mahlon Bush 110
Hodgkin, Dorothy 147
Hofstadter, Robert 146
Holley, Robert 112
holography 143
hominids 131, **132–136**, **138–141**, 153
Homo erectus 140–141, 153
Homo habilis **138–139**, 141
Homo sapiens 141
homosexual transmission of AIDS 123
"hot line" xii
Hou, Ya-Ming 151
Hoyle, Fred 54–55
Hubble, Edwin P. 39, 50, 55
Hubble's Law 39, 50–51
Hubble Space Telescope 41–43, 151
Hudson, Rock 123
human beings, origin of **131–141**
human growth hormone 130
human immunodeficiency virus (HIV) *See* HIV (human
    immunodeficiency virus)
Humason, Milton 39, **50**, *50*
Huntington's disease 129–130
Hutton, James 91
Huxley, Thomas Henry 92
hydrogen 5, 43–44, 75, 80–81, 83–84, 115–118, *116 See
    also* deuterium (heavy hydrogen)
hydrogen bombs (H-bombs) xiv, 25, 144, 147
hyperons 22–24
hypotheses ix–x

I

ice 76, 78, 83
ice caps, polar 97
immune system 123–124
inflationary model of the universe 58, 61, 149
influenza viruses 125
Infrared Astronomical Satellite (IRAS) 40
infrared lasers 17
infrared light 41t, 69
Inoue, T. 143
insulin 130
interferon 149
interleukin 2 130
intravenous drug use 124, 126
Io (moon of Jupiter) 63, 76
ion microscopes 144

iridium 93–94
iron 70, 84
Iron Curtain xi
isotopes 134

J

Jacob, François 111
Jansky, Karl 45
Jeffreys, Alec 150
Johanson, Donald 139–140
Joyce, James 31
J/psi mesons 31–32
Jupiter (planet) 38, 63, 73, **74–77**, 76, 83–84
Jurassic age 94

K

Kamerlingh, Heike 18
kaons (K-mesons) 15, 22–24, 26, 29
Kaposi's sarcoma 123
Keck telescope 39
Keith, Arthur 136–137
Kennedy, John F. xii
Kepler, Johannes xv
Khorana, Har Gobind 112, 147–148
killer T cells 124
Kimeu, Kimoya 140
King, Thomas J. 120
K-mesons *See* kaons (K-mesons)
Korean War xi
Kornberg, Arthur 111
K-T boundary 94
Kuiper, Gerard Peter **66–67**, *67*
Kuiper Airborne Observatory 79

L

LAGEOS (Laser Geodynamics Satellite) *90*, 91
lambdas 15
Landsat I (U.S. satellite) 88
Large Magellanic Cloud 150
lasers **16–17**, 143, 146, 149
Lawrence, Ernest O. 17, 20
Lawrence Berkeley Laboratory 20, 93
Leakey, Jonathan 139
Leakey, Louis Seymour Bazett 132, 134–136, 138
Leakey, Mary 132, 134, 136, 139
Leakey, Meave 139
Leakey, Richard 139–141
Lederburg, Joshua 128
Lederman, Leon 62
Lee, Tsung Dao **24–28**, *25*, 29–30, 145
Lemaître, Georges 54
leptons 7, 32–33t, 143, 153
Leucippus xv, 2, 4–5
leukemia 122
Levene, Phoebus Aaron Theodor 101
Libby, Willard Frank 134, 143
Lick Observatory (California) 66
Lie, Sophus 29
life, theories of origin of
    clay **117–118**
    primordial soup **113–117**
light 10, 41t, 58 *See also* lasers; red shift
lightning 115, *116*
Lin, Chia 147
lithium 20
living floors 133
Los Alamos (New Mexico) 8, 30
Lowell, Percival 71–72, 83

INDEX